Garden Poultry Keeping

J. Barnes

Northbrook Publishing Ltd

Beech Publishing House

Station Yard

Elsted Marsh

Midhurst

West Sussex GU29 OJT

ISBN 1-85736-257-8
First published 1996
Reprinted, 1997
Reprinted 2000

New Edition 2001
with coloured plates.

British Library Cataloguing-in-Publication Data
A catalogue record for this book is available
from the British Library.

Northbrook Publishing Ltd

Beech Publishing House

Station Yard

Elsted Marsh

Midhurst

West Sussex GU29 OJT

CONTENTS

Aseel which came from India
Not very productive, but fascinating to keep.

FOREWORD

This is an introduction to the fascinating hobby of poultry keeping for those who wish to keep a few chickens in the back garden, orchard or paddock. Selection of the most suitable breed is important, but many have special features which make the hobby more interesting so 'shop around' before deciding which standard breed to purchase.

Unlike many hobbies, it is useful as well as interesting. Producing fresh eggs, table birds, manure for the garden and a chance to meet other poultry fanciers is all worth while.

No previous knowledge is assumed so the text covers in detail all the essentials.

Thanks are due to many artists and poultry keepers and authors whose material has been used to illustrate the text.

OPPOSITE

Poultry Breeds which originated in different countries.

These have been bred for keeping in domesticated conditions, unlike the **hybrids** which were genetically modified for keeping in battery cages, a system which is cruel to the hens.

British ORPINGTONS

American WHITE WYANDOTTES

Spanish MINORCAS

Belgian CAMPINES

Italian LEGHORNS

Russian ORLOFFS

Japanese YOKOHAMAS

French HOUDANS

Chinese CROAD LANGSHANS

Dutch BARNEVELDERS

R. FARWIS

Polish Fowl – White crested blacks and Laced

These are quite acceptable layers and are an ancient breed, well
worth keeping, but provide water founts that do not allow the
crests to get wet when drinking.

1
WHY KEEP POULTRY?
Pros and Cons

There is considerable pleasure in keeping poultry for pro ducing eggs for eating direct or for baking; moreover, they are fresh and the keeper knows that good whole some food is being given to them. The many aspects are examined below.

ADVANTAGES

1. A worthwhile hobby which can embrace poultry shows and being a member of a club and attend functions where you can meet people with a similar interest.

2. Fresh eggs are marvellous for all kinds of food dishes so the housewife can use them with confidence. They are high in protein (about 14 per cent). The birds are allowed out and live in the fresh air not cooped up in battery cages.

3. Poultry food can be purchased from the local mill and is usually quite economic and in pellet form is easy to feed. In addition, household scraps, such as bread, fried bacon scraps, and cabbage leaves, can be utilized, although scraps should be limited to not more than 25% of the total. All household waste should be cooked. Leaves, garden weeds, grass clippings and other by-products of the garden provide great food value in the summer.

4. Keeping poultry is an excellent hobby and provides great interest **and** relaxation. Rushing around in a car is one of the most over-rated pastimes, especially on bank holidays; yet with a garden and pen or two of poultry the interest is provided and it pays for itself.

5. Chicks or growing stock may provide a lucrative income, especially if a winning strain is established.

DISADVANTAGES

1. Capital Costs. As would be expected a suitable shed and run is necessary and this can be built or purchased. The cost should be recovered quite quickly. *Poultry Houses & Appliances A DIY Guide* shows how to build sheds and is available from the publisher.

Moreover, properly planned the poultry house can be integrated into the garden design, no matter how small. If an orchard or paddock is available all the better because, in effect, the birds can have free range, where they will forage for food and save at least 10 per cent on food costs.

The author spends many happy, relaxing hours taking care of the birds and, on a warm day, sitting on a bench under the trees, watching the courtship of the cockerels, the rearing of chicks, and the general behaviour of the birds, as they spend their daily lives searching for tit bits amongst the leaves.

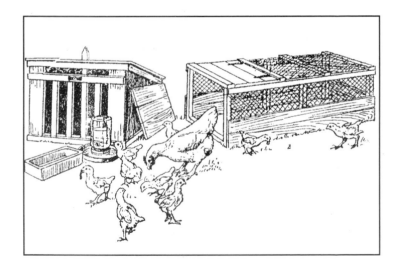

Many Relaxing Periods are Possible Watching the Growing Chicks

2. Possible Restrictions. There may be local authority or other regulations in force which prohibit the keeping of poultry and this must be checked before pens are established. Usually in rural areas there is no problem. Obviously, fowls must *not* be allowed to stray on other people's property.

3. Noise from the crowing of cockerels may be a nuisance to neighbours so take care not to infringe on the quietness of the area or trouble may result. If there are difficulties, remember that a cock is not essential unless breeding is intended. Hens will lay just the same without a male.

4. Holiday arrangements. When on holiday, arrangements will have to be made for birds to be fed and eggs collected. Usually this is not difficult to arrange with payment from eggs collected.

TYPES TO KEEP

There are many possibilities and which approach to adopt depends on the space available and the breed which appeals. Many in the past have kept "hybrids" which are poultry bred for commercial purposes, usually for battery cages.

They are not to be recommended for garden poultry keeping when there are so many interesting varieties available. The term 'poultry' embraces many different types and breeds of bird:

1. Standard breeds of fowl

There are about 60 different breeds and their varieties which are recognized by various poultry clubs. They are divided into **Heavy** breeds (sitters and weighing from about 7lb. (3.20k.) to over 13lb. (6.00k +) and the *Light* breeds – non-sitters up to about 6/7 lb. There is some overlap.

2. Bantams

A similar number to large fowl exist and these are miniatures of the large breeds, or natural bantams. They are about a quarter of

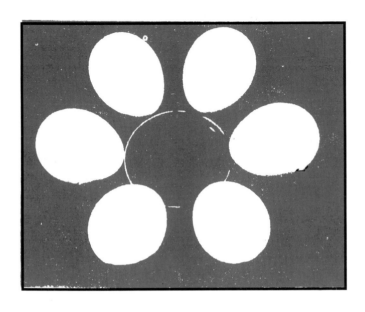

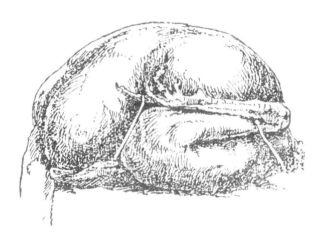

Fresh Eggs can be provided as well as Table Birds

the size of large fowl and lay eggs about half the normal egg (35g+ compared with 60g+). About 150 per year is the possible output from laying strains; some lay 250!

3. Ducks and Geese

Normally these are kept where there is an orchard or paddock and, preferably, a pond because too close confinement is not good for them. A few ducks may be kept in a water garden with fibre glass ponds (or other types) and running water and this can be very attractive. However, a pond is *not* essential.

4. Guinea Fowl

Very interesting birds, that can be noisy. Act as watch dogs' making a loud cry when visitors arrive. Very thick shelled eggs.

5. Peafowl

If the garden is quite large what better than a magnificent pair of peafowl. Again, they are noisy and require a large space.

6. Turkeys

These can be kept in a large garden and fattened with semi-intensive facilities or in an open space for foraging. There are many ornamental types of turkeys such as Buffs, Bourbon Reds, Blues, and even one with the delightful name of Royal Palm.

SCOPE OF THIS BOOK

The concern in this book is with the standard breeds of poultry. What is stated also applies to bantams, the main difference being that they will require about half the space normally allowed for large fowl. Possible further reading from the publishers:

Keeping Guinea Fowl,	*Bantams & Small Poultry,*
Ducks, Geese & Turkeys,	*Keeping Peafowl.*

Attractive Features can be Made as well as Unusual Designs for Equipment.

The barrel above includes automatic feeding and water supply designed with a rustic appeal. The water is held in a metal or fibre glass tank with a pipe leading to the outside water trough. This may be controlled by a small ballcock or other device. The food is in a separate, dry compartment and relies on the hens pecking to release the corn which is operated by gravity feed.

The original version was operated by the chains being pulled to release corn and water and was kept open for 30 minutes morning and evening.

2

ACCOMMODATION

Arrange the Housing First

Before making a decision to embark on poultry keeping make sure that adequate housing, including: a run, is available. Satisfactory results can only come from purchasing or making a suitable shed or modifying a garden or tool shed.

The house should be established with the following rules being followed:

I. Size

Adequate for number of birds to be kept. A shed 1.5 metres **x** 2 metres should be adequate for about six hens, and up to 12 if these is a large run. However, do not overcrowd. A smaller shed would house on a pro-rata basis.

2. Srong and Vermin Proof

Besides strong enough to keep out the elements and should *also* be rat and fox proof. Usually these pests can be kept at bay, but should be safeguarded against and not encouraged by food being left around.

A strong floor, preferably of concrete or, if off the ground, on stout sleepers; thick, creosoted board, should be used.

3. Ventilation Adequate without Draughts

Poultry cannot stand cold, damp conditions or excessive draughts so care should he taken to put vents which do not blow directly on to the birds.

4. Pleasing Design

The experts usually specify that the design should have *aesthetic appeal* (pleasing to the eye) and be *practical.* These two requirements apply with great force with a garden poultry set-up. Nothing looks worse than a home made shed with rusting corrugated iron sheets and patchwork building. A badly planned layout stuck in the middle of a garden is an absolute eyesore and attracts vermin.

5. Screens and/or a Special Section

The fencing used can keep the birds within their domain and also landscape the area so that the poultry unit enhances the general appeal of the garden. Trellis work, interwoven fencing, link fencing or other dividing material, interspersed with bushes, would be ideal. There could be a 'no-man's area' to separate the poultry unit completely and this should be located at the bottom of the garden.

EXAMPLE OF SPECIAL AREA

The possible area for the poultry shed and run, and plan, could be as illustrated below and opposite. This shows the fence being erected and the dividing area between the garden proper and the poultry.

Layout of end-of-garden Layout Poultry Area.

Small shrubs, climbing plants on trellis work, and attractive layout which makes the area an integral part of the garden should be planned.

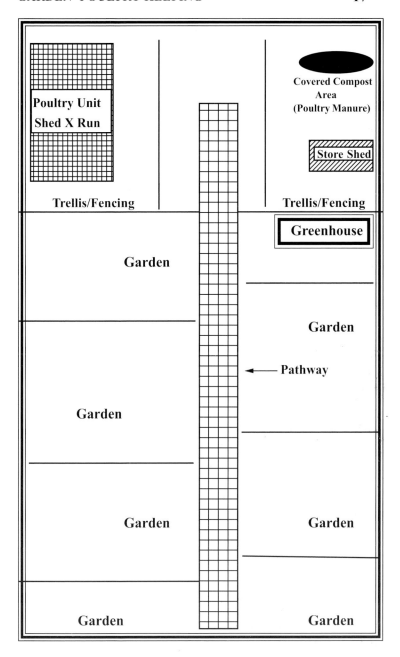

Plan of Possible Garden Layout (See Illustration opposite)

Possible Types of Building

There are very many types of poultry house which could fit into the garden layout. If an orchard or largish area of grass is available some kind of movable ark can be used, thus giving the birds the benefit of fresh grass which makes excellent food.

If space is limited the house will have to be relatively small with a run combined with provision for this to be covered in winter or it will become water logged. Some types of accommodation combine sleeping quarters and a run for scratching, all under cover. In that case a regular supply of greenstuff will have to be supplied to the birds as well as floor covering such as leaves or shavings. Weeds from the garden with soil attached makes good foraging material.

The possible types of houses are as follows:

1. Small House with outside run which is simply an area of garden fenced off and the birds are closed in each evening (a vital precaution against predators).

2. House with internal run. There are many clever designs available.

3. Sussex Ark. A popular shed for those who have a patch of grass.

4. Fold Unit which is a pen with a covered area at the end for sleeping and laying.

NOTE: There are many other variations which can be purchased or made. Detailed specifications of houses are to be found in *Poultry Houses & Appliances (A DIY Guide)* from the publishers.

Sun Porch/Run
with wire floor
for easy cleaning

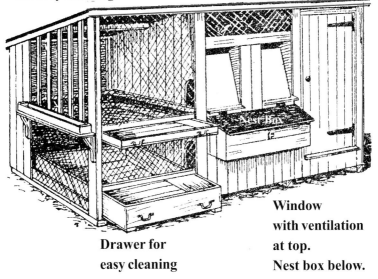

Drawer for
easy cleaning

Window
with ventilation
at top.
Nest box below.

Small Poultry House & Run

This requires very little attention, being self contained. The food can be in a hopper inside. Ideal for bantams. The house allows the birds to stay in all day when the weather is inclement or there is nobody at home. Obviously, this can also be used with an attached run.

Picturesque Shed with Thatched Roof
Wire front provides fresh air, but may be varied to suit weather.

Slatted Floor House
On wheels so can be moved around and is easy to keep clean;
droppings fall below for easy removal.

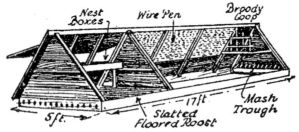

Fold Unit
Can be moved each week on to fresh grass.
Make sure the floor is wire meshed to stop burrowing
vermin.

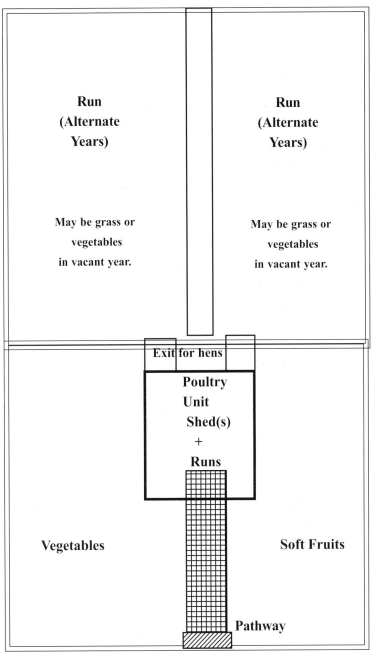

Run
(Alternate
Years)

May be grass or
vegetables
in vacant year.

Run
(Alternate
Years)

May be grass or
vegetables
in vacant year.

Exit for hens

Poultry
Unit
Shed(s)
+
Runs

Vegetables

Soft Fruits

Pathway

Alternative Layout for garden or allotment

A
Lean-to
Intensive
House.

A Span
Roof"
House
with
"Top
Lights".

CANVAS

A Movable House .

A Backgarden
House.

A
Tolman"Long Slope House"

A Sussex
"Ark"

A Variety of Sheds for Small Poultry

It is important not to overcrowd.

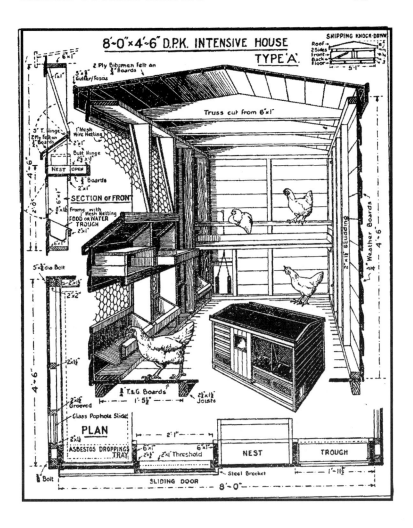

Model Poultry House

This includes measurements and the essential equipment so that an enthusiastic amateur joiner could build all that is required. The overall result is shown by the shed in the centre of the illustration.

The Choice

Once the possibilities are examined the appropriate shed and run can be made ready. Do not overcrowd, especially if there is limited scratching space. There should be adequate perch space and room for the food and water hoppers. If pellets are to be fed they should be kept under cover.

As a general rule, each bird should have 2 sq. ft. of space in a shed with 3 sq. ft in an outside run. A typical shed (1m x lm) plus a run of 2m x 1m should be adequate for six to eight layers, but ensure there is adequate perching space. When there is no outside run each bird should have 5 sq.ft of space. The height of the shed should be 6ft (2m), for each bird will require 25 cubic ft of breathing space. Fresh air is vital !

A shed should be of the type that can be erected in sections, each bolting together. This facilitates removal and repair. *Remember that birds can make the ground sour and therefore a move every two or more years will be advisable. Each year the outside run should be turned over and limed.* If leaves and grass clippings are added regularly the run will be kept fresher and the additions are a source of food. A soil which drains well is to be recommended, or the birds will very quickly get bogged down with mud. If birds are to be let out in winter in an open run, try to have at least basic draining channels and, near the pophole or exit, it is helpful to place a thick layer of shingle or pebbles so that mud is not carried in, which will soil the floor and dirty the eggs. Low, wooden staging can also be used.

Storage Bins

Food should be stored in a steel hopper. An old, disconnected chest freezer makes an excellent storage unit and this should be placed in a suitable shed.

Try to avoid leaving food around or rats will be attracted. At the first sign of these creatures (evidenced by sight or droppings) put rat poison in a container with a hole for the rats to go into, not accessible to the birds.

Old Style Poultry Yard

Ideal when keeping a number of breeds when showing.

SHOW REQUIREMENTS

Once poultry keepers are established and become interested in showing at local or even National shows it will be necessary to have a number of pens and a Penning Room for training the birds for showing. A layout is given on the preceding page. Make sure the bottom part of the pens are boarded to avoid cocks sparring.

Showing can be very interesting and rewarding and does make sure the birds are kept up to the correct standard. The birds have to be tamed in training pens and also washed, but this is beneficial becuse it helps to keep the birds clean and free from parasites.

Handling the birds correctly is also important. Never carry a bird by its legs or throw it on the ground. Once tamed they will eat out of your hand and therefore picking them up indoors is no problem. In an open run birds will naturally fly away, but a catching net can be used, which avoids birds getting alarmed.

Typical Layout for Keeping Small Stud

May also be used as cockerel pens before deciding which birds to keep. The above is intended for somebody who has established a number of breeding pens, and approaches the small-holding level. For those keeping a few birds this much will not be necessary.

WASHING

The need to wash birds must be acknowledged because they should never win if dirty or in poor condition. In any event, washing reveals any defects that may exist and are not apparent from a superficial examination. The points to watch are as follows:

1. **Wash about 4 days before the show so the feathers are fully dry before exposing a bird outside and in the show. Feathers which are not dry never hang properly and therefore the bird in question is at a disadvantage.**

2. **Use warm water at a temperature which allows the hands to be immersed, yet warm enough to fetch off the dirt.**

3. **Most fanciers use two bowls, some three, to wash thoroughly and then rinse in one or more lots of clean water at a temperature which is just warm. The first bowl should contain washing-up liquid or a hair shampoo, and later bowls are for rinsing and adding any conditioner. The cleanser must not be too strong or it will take too much of the natural oils from the feathers.**

4. **Wash the legs first, outside the main bowl — usually on the side of a sink so the dirt can be washed away and not go in the washing bowl. Use a medium stiff scrubbing brush to remove all the dirt.**

Drying

The birds should be dried in a warm room, but not too hot. Various methods are used, including a drying cabinet rather like a hospital cage with built-in heater, *or* a small hair dryer, or in front of a fire or radiator in clean, straw-filled baskets.

After drying the birds are put back in the show pens and kept as clean as possible and immediately before the show the legs are washed again and polished with a cloth. Remember no faking as such is allowed, although it must be appreciated that the birds must be turned out to the best advantage.

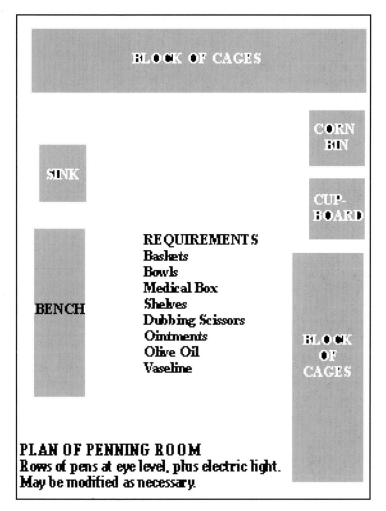

BLOCK OF CAGES

CORN BIN

SINK

CUP-BOARD

BENCH

REQUIREMENTS
Baskets
Bowls
Medical Box
Shelves
Dubbing Scissors
Ointments
Olive Oil
Vaseline

BLOCK OF CAGES

PLAN OF PENNING ROOM
Rows of pens at eye level, plus electric light.
May be modified as necessary.

**Penning Room for Storing Food and
Penning Birds for Training.**

This is not essential, but can be extremely useful and can double up as a potting or tool shed.

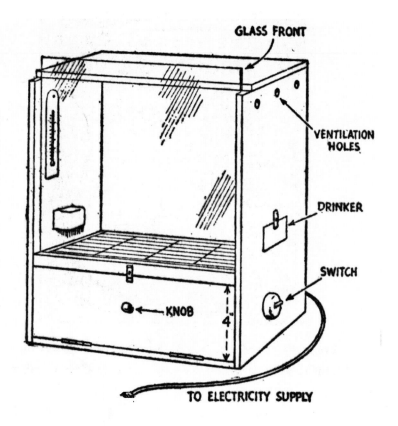

GLASS FRONT

VENTILATION HOLES

DRINKER

SWITCH

KNOB

TO ELECTRICITY SUPPLY

Drying or Hospital Cage

This is a useful addition when birds have been washed. Set at a warm temperature and a bird very quickly dries after being washed as described earlier.

An alternative is to use a hair dryer with the bird in a basket to keep it still.

If a bird has been injured it may be necessary to place it in a warm location until it feels better, when it can be kept in the penning room in a normal cage. An alternative is to place in a small show basket with straw in the bottom.

Basket Which May be Used for taking Birds to Shows.
May also serve as a drying basket or temporary hospital place.

Basket for Bantams; one bantam per compartment

Single | Communal

Nest Boxes

These are a vital part of management. On the left is a block of single boxes which give more privacy to the hen. Note the gangway in front for ease of access. The sloping top prevents the birds from perching on the nest boxes on a night.

The communal boxes are more economic to make, but could be more trouble with quarrels occurring.

Each week the shavings or straw should be renewed. Spray inside when birds are outside to kill mite. Remove eggs at least once per day. Take broodies off the nest each evening.

Do not encourage birds to lay on the floor. Avoid dark areas in the house because hens will lay there and very quickly come broody when eggs accumulate.

A Basket Nest Box
This may be moved around and can be attached to the wall if required.
If a hen comes broody can be moved to a separate house for privacy.

Top: Anconas: Fine layers of white eggs

Bottom: White Leghorns. Another Mediterranean breed; good layers and white eggs. Not as tame as heavier breeds.

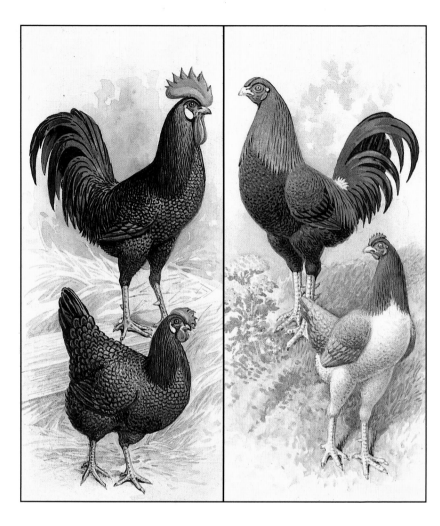

Andalusians: Sound layers of white eggs. Do not come broody. Quite scarce, but well worth keeping for their beauty and laying. Too small for table birds. May be on the wild side for small garden.

Old English Game: Black Red Cock and Wheaten hen. Rather dark wheaten by today's standards–more like the 'Clay' colour. Reasonable layers and fine table birds.

Light Brahmas: Large, exotic birds; moderate layers of tinted eggs. Feathered legs and Pea comb make them very attractive. Heavy breed which will not fly over fences or become wild in behaviour.

Dark Brahmas. Important because they were the breed used to create others. The dark colour helps to keep them clean.

Top: Campines. These days a Hennie type of cock is preferred (with no sickles). Lays about 150 white eggs a year.

Bottom: Dorkings – Silver cock and hens, and dark hen. Heavy bird, but no longer very productive. Five toes.

Buff Cochin: Only fair layers. Very feathery and majestic. Not good layers; tinted eggs.

Partridge Cochin: Very attractive colour, but not a commercial type bird. Single combs, whereas Brahmas have a Pea Comb.

Top: White Dorkings which have Rose combs, and Red Dorkings which are quite rare. One of the oldest breeds, but have lost ground.

Bottom: Faverolles: French breed for laying and table. Tinted eggs in reasonable numbers. The colour shown is really a non-standard variety because the nearest is the Salmon which is similar to the above, but has shoulders of cherry red and gold (USA Reddish brown). Note the muffles, five toes and feathered legs. Weight of cock about 5 kilos (11lb.).

Top: Gold Pencilled Hamburghs. Once very popular and quite beautiful.

Bottom: Silver Spangled Hamburghs. Spangles and tips make this a very attractive breed. Light Breed.
Both lay white eggs in reasonable quantities, but on the small size.

Top: Houdans. Attractive with crest and leaf comb, but only fair layer.

Bottom: La Fleche. French like Houdan. Lays a white egg and is a good table bird, but rarely seen. Unusual horn comb.

Croad Langshan. Good layers of deep brown eggs as well as being commendable table birds. Probably source of deep brown eggs for others dark brown egg layers. ('Coffee' Coloured)

Modern Langshan. Taller than *Croad* on which it was based. Brilliant black plumage, but now rare.

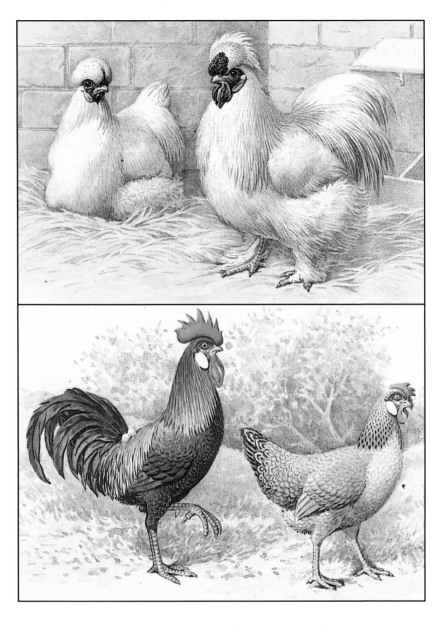

Top: White Silkies. Very popular breed and a renowned broody. Become very tame. A good layer when not broody, but not a table bird.

Bottom: Brown Leghorns. Very attractive member of white egg layers.

Top: **White Minorcas: Delightful member of Minorca family. Many prefer Black variety because they keep cleaner looking. Non-sitter. Quite rare along with Blues.**

Bottom: **Black Minorcas which are great favourites.**

Top: Buff Orpingtons. A great favourite in days gone by. Dual purpose breed for laying and for the table. Laying abilities of some strains rather poor because they are now too feathery.

Bottom: Blue Orpingtons. Original type as developed by William Cook. Very popular and quite attractive. Large birds for table.

Top: Barred Plymouth Rocks. A popular dual purpose, heavy breed which lays well (tinted eggs). Very old breed which settles well on free range or in a small area.

Bottom: White Wyandottes. Like the Rocks is an American breed which lays well. Heavy breed which lays tinted eggs. Suitable for free range or semi-intensive in open run.

Top: Dumpies or Scots Dumpies. Dwarf-like breed which lays reasonably well, but belongs to a special race known as "Creepers" which, due to the genetic makeup may carry a lethal gene. Lays white eggs. Light breed.

Bottom: Scots Grey: A gamey type breed which is barred. Good layer of white eggs and quite attractive, although not many are kept. Non-sitter.

Red Sussex. Dual purpose breed which lays light brown eggs. Fattens well for the table. Mature cock over 4 kilos (about 9lb.) This variety rare.

Light Sussex. These are a utility type. They are very popular, but for showing the hackle and tail must have very dark markings.

Various Breeds
Barnevelders, Marans, Welsummers (Deep brown eggs)
Rhode Island Reds (Brown eggs), White Leghorns (White eggs),
Indian Game (USA Cornish) Table bird, especially when
crossed with white skinned breed.

3
OTHER REQUIREMENTS

Vital Equipment

Once the shed and run are ready the birds should be obtained and placed in the shed. For about a week, feed and water them indoors and then, once they are settled, they can be let out into the run. For the first few days of their release outside into the garden or an orchard, they may have to be driven into the shed until they become used to going in.

An alternative is to have a run attached to the shed and allow the birds into that enclosed area where they are kept quite safe and, most importantly, they have a plentiful supply of fresh air, and space in which to scratch.

The vital equipment for keeping the birds will be as follows:

1. **Food Hopper**
2. **Water Container or Fount**
3. **Nest Boxes (see *Opposite*)**
4. **Perches**
5. **Droppings Board or Tray.**
6. **Grit Hopper.**
7. **Greens' Rack.**
8. **Dust Bath.**

Food Hoppers

There are many ways of feeding the birds and the best for the keeper should be selected. Remember that the food must have adequate vitamins, protein and edible roughage. Wheat by itself is not enough and this is why many adopt the easy way and have hoppers which allow birds to eat when they want. which is self-adjusting.

We prefer the pellets because they are more palatable and the 'mash', which is dry meal, sticks to the beak and goes into the water. The evening meal may be mixed corn scattered in the litter so the birds will scratch. There are various sizes available so counting on 4-5 oz per day per bird, place enough in the hopper to cover a week. ***The exact quantity depends on size of birds and quality of food***. Do not over fill or the food will go mouldy.

Water Containers

These can be bowls or other containers, but the most suitable is the water fount which regulates the flow of water and prevents dirt getting in the main container. A two-gallon (10 litre) fount is suitable for adult birds and this should be checked daily. Immediately the water runs out wash the utensil and replenish. with clean, fresh water. Do not take from a pond or container which contains stagnant water. If water is missed, even for a day, the birds will suffer and may resort to eating their eggs, a habit which must be avoided.

Nest Boxes

Nest Boxes should be provided for they allow for privacy and are normally quite clean and should be sprinkled with insect powder. Once every 6 months, spray in cracks where mite may hide. Two compartments will suffice for up to 6 hens and a 'curtain' of some form of hessian will provide privacy.

The nest boxes may be lined with hay, straw or shavings; this should be changed regularly or the eggs will get soiled.

Perches & Droppings Board

Poultry like to fly up as high as possible and roost and for that reason there should be adequate perches for them all to find a place. About 1 foot (30cm) per bird is the minimum requirement for layers and heavier birds will require more.

The perches are best over a droppings board which is a shelf designed to catch the droppings; it should be covered in litter and the droppings raked off each week. The ideal perch width is 2 inches (5 cm). and the top side should be rounded off.

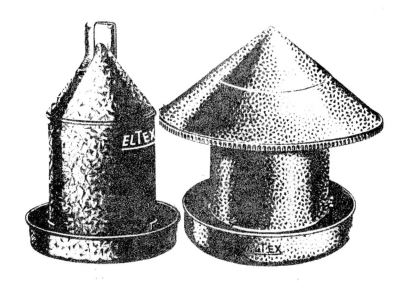

Drinking Fountain
Top pulls off for easy
cleaning

Food Hopper
Suitable for outdoor
use.

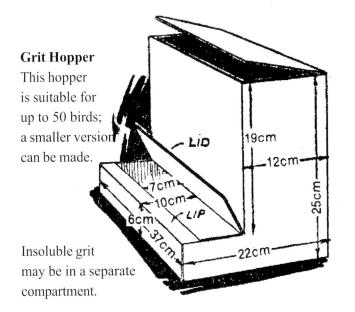

Grit Hopper
This hopper
is suitable for
up to 50 birds;
a smaller version
can be made.

LID

19cm

12cm

7cm

10cm

LIP

6cm

37cm

25cm

22cm

Insoluble grit
may be in a separate
compartment.

For ease of cleaning the perches should be placed in slots, U- shaped. When heavy breeds are kept make sure the perches are not too high or the birds may injure themselves. Cocks of the heavy breeds cannot fly up more than a short distance so the perches should be just above the ground. Stepped perches may also be useful to allow the birds to reach a main perch which is about 1 metre high or more, but lower for really large birds.

If natural branches are available cut a suitable length of the appropriate thickness and suspend above the droppings board. The droppings board keeps the faeces quite separate from the floor litter and thus gives more space; as a result the arrangement is more hygienic. The droppings also dry more when above the ground provide valuable poultry manure, which should be dried before being used. If the droppings board is also removable this may be taken outside and washed down. It should be dusted with dustbin powder or washed with a mixture of water and Jeyes Fluid.

Grit Hopper

Grit is essential in two forms:

1. Soluble Grit such as oyster shell or lime-stone which provides calcium for the well- being of the bird and a coating to form the egg shells.

2. Grinding Grit: Small pieces of flint grit to grind up the food when it reaches the Gizzard, which is a kind of organ which the bird uses to make the food digestible.

The hopper should be kept under cover so that the rain will not sour the soluble grit. The insoluble grit– the small flints—will be scattered in the run so the birds can pick it up when required. When layers' pellets are purchased from the local mill they will contain limestone or similar grit so some is being provided. However, because corn may also be fed it is wise to have other extra grit available at all times. Birds running outside and foraging will pick up grit and earth from the run.

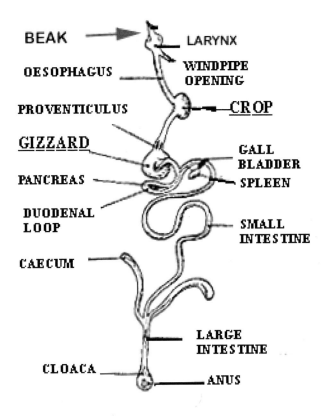

BEAK → LARYNX

OESOPHAGUS

WINDPIPE OPENING

PROVENTICULUS

CROP

GIZZARD

GALL BLADDER

PANCREAS

SPLEEN

DUODENAL LOOP

SMALL INTESTINE

CAECUM

LARGE INTESTINE

CLOACA

ANUS

Digestive System of the Fowl

Food is taken up by the beak and then passes through the system. After the Crop the food passes to the Gizzard where it is ground down by muscular action and the small pieces of flint which the bird must have for digestion.

Pellets and mash from the food millers will contain the soluble grit to the correct level, but some should still be available in a hopper. When birds are running outside they will pick up much of their requirements.

Greens & Greens' Rack

A regular supply of greens is vital and therefore any birds which are kept in a closed run should be fed greens at least every other day. A rack may be provided or a cabbage may be hung from a post and the birds peck this as required. However, do not hang too high so the birds have to jump. Hens are not gymnasts and may injure themselves if the greens are too high.

Sprouting greens may also be used to supply the vital greenstuff. At one time there were special, metal cabinets available on which trays could be used to sprout corn. Within a few days, in a warm temperature, the corn sprouts and grows.

It has to be watered each day and a nutrient solution can be added to encourage growth. The plastic tray is then put into the pen and the birds will eat every morsel.

Even simply soaking grain in a bucket of water for 24 hours and then draining and leaving the grain to sprout can provide an ideal food which the birds eat very readily. It will help the growth if a sacking is placed over the tray of sprouting corn. In two to three days the corn will be ready to be consumed.

Mixed corn, including maize, is recommended because this gives a better vitamin coverage (Vitamins B and C) than wheat alone.

Dust Bath

Unfortunately, the warmth of the body of the fowl does attract scavengers such as lice and mite. If these are to be controlled then the hens should be supplied with an area in which dry, fine soil is available in which the hen can dust herself. If space is limited then a box about 2 foot square (60cm^2) can be supplied in which is placed fine soil, sand and fine ash. If a really dry mixture, a hen will throw this up onto her back and roll in the dust. This has a cleansing effect and clears the parasites.

An additional precaution is to pick up a bird and check around the vent; sulphur ointment or vaseline with inclusion of a mild disinfectant very quickly clears the lice. Never rub with paraffin or other chemicals which will burn the skin.

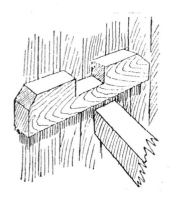 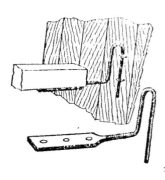

Perches Slot in to Groove.   **Alternatively Hook-on method**

The hook method avoids mite collecting in the groove; the latter
should be creosoted to kill any vermin.

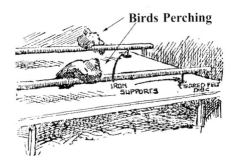

Perches & Droppings Board

Perches should be wide enough to grip and slightly rounded. The top fittings are for supporting the perches. The bottom drawing shows perches and droppings board with birds on the perches. Shavings may be placed on the droppings board or, alternatively, if the manure is to be dried and used on the garden, it would be better to use peat moss which is dry and which soaks up any moisture.

**Dusting
with
Insect
Powder**

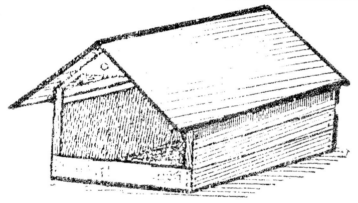

Above: **Dusting a Bird with Insect Powder**
Bottom: **An Outdoor Dust Bath**
for the birds to clean themselves.

Cleanliness

Poultry are quite clean provided they are not neglected. They must be given the means for cleaning themselves as well as regular cleaning out of the shed and dusting where necessary.

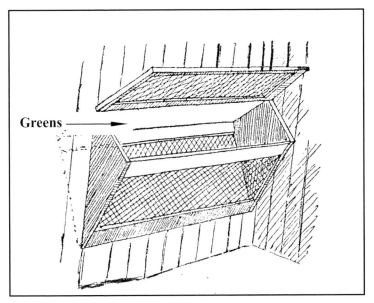

Above: **Rack for holding grass, cabbages, leaves, etc.**

Turnip (Swede)
Being grated for
the birds when
greens are scarce.

Stand for Hanging Cabbages.
Do not place too high for birds
should not be made to jump very high.

Black Minorca Hen   White Leghorn Hen

Ancona Pair (Utility type).

**Mediterranean Breeds which Lay White Eggs in large
numbers.**

These breeds do not generally become broody. The hens weigh
about 6/7 lb (about 3.5 k) and are not good table birds. May not
be suitable for a small garden with near neighbours because they
can fly, but ideal in a covered run with a roof or wire netting.

4

SELECTING THE STOCK

Selecting the Breeds*

Thehe questions which must be answered before starting to keep poultry are referred to in business as *determining the objectives* and would be along the following lines:

Which is the Appropriate breed to Keep ?

The correct answer will depend on a number of factors, the main ones being as follows:

1. Egg Colour:

Are brown egg- *or* white egg layers required;

2. Egg Target:

Is the target for maximum eggs to be a criterion (will affect choice);

3. Location and Noise:

Can 'flighty' birds be kept or is there so much noise that placid birds will be better;

4. Exhibition Birds:

Is there a possibility that birds will be exhibited;

5. Availability of Breeds:

What breeds are readily available which have been seen at shows or are being advertised? Avoid 'hybrids' because these are unsuitable for the garden having being developed for cages.

---

* **For a more detailed description of the breeds readers are advised to study *Breeds of Poultry and their Characteristics*, Batty, J (from the publisher)**

Once a satisfactory answer has been given the search for suitable birds can start. The possibilities are listed below:

1. Dark Brown Egg Layers
Marans, Barnevelders, and Welsummers.

These birds lay very dark brown eggs, but their output is not likely to be as good as some of the other breeds.

2. Brown or Tinted Eggs

Many birds lay a lighter brown egg or tinted (creamy) colour, and these include such breeds as:

Australorp, Brahma, Cochin, Croad Langshan, Dorking, Faverolles, Jersey Giant, New Hampshire Reds, Orpington and Rhode Island Red.

3. White Egg Layers

These include a variety of breeds, the most important being the Mediterranean breeds **Anconas, Leghorns** and **Minorcas.**

There are other, more exotic breeds such as **Polands, Redcaps, Hamburghs, Old English Pheasant Fowl,** and **Old English Game** which lay a whitish egg, although sometimes these tend to be a creamy white.

The Mediterranean fowl are splendid layers (300 eggs per annum have been recorded at the top end), but they do tend to be rather flighty and if they are to be kept, there should be wire netting or some form of nylon netting across the top of the pen or they may fly and escape.

4. Dual Purpose Table Breeds

If birds are to be killed for the table it is necessary to select a breed which fattens very well and if laying capacity is also essential, to select breeds which are dual purpose. **Indian Game** are the main table bird, but develop quite slowly, and dual purpose birds include **Dorkings, Sussex, Orpingtons** and **Wyandottes.** Sometimes Indian Game are crossed with another breed, such as Sussex, which have white skin and this cross produces excellent table birds.

White Wyandottes   Barred Plymouth Rocks

Faverolles

Buff Orpington Hen

Dual Purpose Breeds
Not to scale

Select Healthy Birds

The choice having been made it will be necessary to obtain healthy stock of the required type.

The appropriate number of birds should also be decided. A normal, laying bird will lay two or three eggs per week and therefore around 6 birds should be sufficient for an average family. In the summer months a Leghorn will probably lay 6 eggs per week so there may be a glut of eggs.

When purchasing try to obtain the following:

1. Young birds just about to start laying

Usually this will be about 6/7 months of age and at this stage the birds will look healthy and their combs should be starting to redden and they will make a noise usually referred to as 'prating' which is a constant cry, usually with rapid movement of the bird. It is best to hatch early in the year or purchase so they start in October for Winter laying. If much younger birds are obtained they will have to be reared in a fold unit and fed growers' pellets

2. Make Sure the Birds are Healthy

Old unhealthy birds should be rejected. Any which look mopey, have *scaly legs*, shrunken, light coloured face and comb, lack vitality or are diseased, should be avoided for they are likely to be no use.

Handle each bird, which to be acceptable, should be plump and firm (not flabby) and have a full coat of glossy feathers, but not too fat. Check the breast bone has no indentations.

Examine the eyes and nostrils and if there *is* any discharge; be careful because this may indicate 'Roup' or bronchial troubles. If the condition is also accompanied by a putrid smell from the nostrils, the birds should not be purchased.

Examine the vent (rectum) to make sure the birds are not infested with lice or have sores. Any lumps or signs of bleeding should also be regarded as a bad sign.

Firm Comb, Satin Texture.
Red Face, Bright eye,
Lean and Active.
Worn Plumage

Carriage
Alert
& Active

Vent Moist
& White

Lean Legs
Little Colour

Typical Light Breed (Leghorn) and its Requirements

Full Bodied weighing around 3 kilos or more.
Good layers of brown eggs, but do come broody,
often at inconvenient times.

Heavy Breeds (Light Sussex and Rhode Island Red) .
These hens are excellent layers in large and bantams.

Defects such as twisted beak, crooked toes, legs which are marked or disfigured should mean rejection. Moreover, birds which are not typical of the breed in terms of weight or *standard description,* (See *Breeds of Poultry & Their Characteristics,* BPH),   should not be kept.

Settling Down

Once the selection has been made the birds can be placed in the house and after a week or so allowed out in a run. Do not let outside too early (eg, on to a lawn) or they will have difficulty in settling down and going back into the house.

If breeding is to be attempted then introduce the cock at the same time, and allowing at least 10 days for eggs to become fertile.

Not all hens come broody and, therefore, make sure the means of hatching is available either from a broody hen or an incubator. The Mediterranean breeds, which are classified as Light breeds, do not usually come broody and sit and this applies to most other light breeds, but the medium and heavy breeds (eg; Rhode Island Reds, Sussex and Wyandottes) do come broody. An important exception to the rule are the light breed Old English Game which hatch well and are generally good mothers, although at the early stages, because of their great energy, can tire the chicks.

Brahma Hen.
Although not
a good layer
is quite
magnificent.

5

BREEDING

The Breeding Pen

As noted earlier, a *male* bird is not essential when no breeding is intended. However, if the poultry keeper wishes to obtain replacements, or if showing, and needs to breed winners, then it will be essential to have a breeding pen.

The cockerel or cock (over a year old) selected should be active and full, of vigour. Generally, in the breeding season, he should also be attentive to his female companions and scratch and find food for them..

Remember, the male is very important; he is half the pen and if his background is doubtful, many of his faults will be transferred to his offspring. This is why some breeders of show birds advocate 'line breeding' or even 'inbreeding' so that related stock are bred together, thus locking in all he desirable characteristics. Thus a son may be matched to the mother, perpetuate the breed show features. . However, for normal purposes , when maximum egg laying is the desirable, the male also should be bred from birds which have a good record for laying.

Unfortunately, in the desire to win prizes many breeds have been spoilt by the owners aiming for features which destroy laying and table potential. Too much feathering, an excessively large comb, or trying to breed a colour which is unnatural for the breed, are all examples where the utility aspects have been lost.

Mating the Cockerels

Usually up to 10 hens to one cockerel is permissible for active birds, but three or four is better in the Spring or when heavy breeds are involved. Chicks should *generally* be hatched in the period February to the end of April., but see below.

If winter layers are to be obtained for medium and heavy breeds the chicks should be hatched in March. These would include Orpingtons, Wyandottes, Plymouth Rocks, Sussex and Wyandottes. These breeds also make the best broodies.

The lighter breeds, Anconas, Campines, Andalusians and Leghorns, should now be mated with a view to hatching in April. These grow quicker than the heavy breeds and will also come into lay in the Autumn and lay through the Winter.

For **show birds** the breeding pen may include a cockerel and two hens (referred to as a breeding trio) and in extreme cases it will be one to one, possibly the son to the mother or the sister to the brother, a process known as "Inbreeding". However, if too closely related vigour may be lost.

As noted, early in the year, fewer hens should be run with a cockerel – until he is fully active and there is sunshine. Also remember that, all things being equal, the heavy breeds can manage fewer hens than the light breeds.

Select breeding stock which is Healthy
Pair of excellent White Wyandottes

Both sexes should be healthy and remember the male represents a very important part of the breeding pen. If he does not have the potential the chicks are not likely to be productive.

Face, Lean & Hollow

Eye Large, Bright & Prominent

Fine, Bright Red Comb & Wattles

Beak Short, Stout & Worn

Head of Ideal Layer
Not all hens have the turned comb

Good sized active Hen Tight, soiled and worn Plumage—Keep.

Select birds as *Above*
Avoid Unproductive Birds (Below')

Ruffled feathers

low tail carriage

loose feathers

cow hocks

Small Bird, sleek, coloured new feathers bright Legs —Cull.

Overfat and Scaly Legs—Cull.

Selection of Healthy Birds is Essential.

The process of elimination of unproductive stock or those birds which are suffering from disease is known as ***Culling*** and is an essential part of good poultry management.

When birds are running on free range as many as 30 hens may be run with two cockerels, but make sure they are compatible. Never run two Game cocks together or they will fight.

Collecting & Storing Eggs

Eggs should be collected at least once per *day* and if possible they should be clean when taken from the nest box. If not clean they will have to be washed on the same day, before any germs have accumulated on the egg.

If for eating they would be stored in a refrigerator. **For hatching** they should be stored in a cool temperature on a tray or box containing shavings. The eggs should be marked with the date and the pen number. When washed a mild disinfectant should be used, thus giving them protection. Do not include disinfectant if intended for eating.

Each day they should be turned and not later than 7 days after laying batches should be placed in an incubator and hatched. Alternatively, a broody hen may be used. Regretfully, early in the season, hens may not come broody and therefore the use of an incubator becomes essential. Remember that the incubator is made to run *at least* half full, so a reasonable number of eggs should be obtained before setting.

Normal Eggs

For all purposes the aim should be to produce good quality eggs with strong shells and normal in shape. Eggs which are quite round, very long or mishapen should not be tolerated, because this is a sign of some defect or fault in the management or the stock. A sound egg will have a shine, which is known as 'bloom' and there should be no faults such as obvious pore holes or what appear to be encrustations on the shell.

Egg shapes differ as do the sizes. The breed and strain have an effect on these factors. Generally it is unwise to bring on a pullet into lay too early or the egg size will suffer on a permanent basis. About 6 months of age is the earliest.

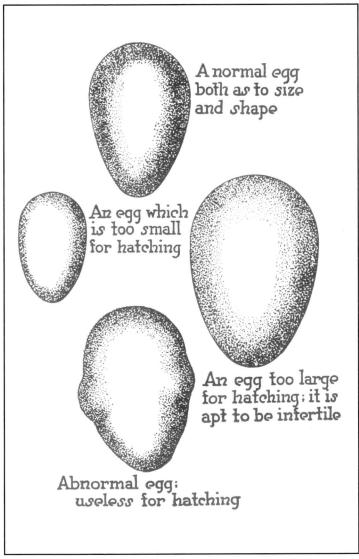

A normal egg both as to size and shape

An egg which is too small for hatching

An egg too large for hatching; it is apt to be infertile

Abnormal egg; useless for hatching

Eggs

Top: Normal egg which should be the aim.

Remainder: Eggs which can be used for eating, but should not be perpetuated by breeding from. Problem should be eliminated. A 60+ gramme egg should be the aim; small eggs are unprofitable.

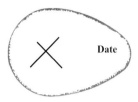

Marking with
X and Y indicates
side turned properly.

Date hatched
& Code shows
Pen Source.

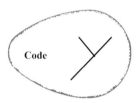

Marking Eggs for Ease of Turning

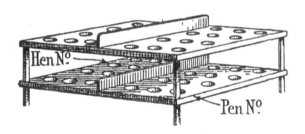

Egg Trays for Keeping Eggs

This is an alternative method to placing eggs in a box with a sawdust bedding.

The Pen and Hen Number can be shown on the trays so that eggs are watched carefully and turned each day before incubation at 7 days or earlier.

WHETHER TO BREED

Whether to breed your own replacement stock is a difficult decision which must be considered in the light of the facts. These may be as follows:

1. **Cock required and this crows and may disturb neighbours.**

2. **Chicks give much pleasure, but are a fair amount of trouble.**

3. **An alternative is to purchase point of lay pullets, large or bantams, but these may not be the breed you fancy.**

4. **If showing is intended then it will usually be essential for breeding to take place because this is the only way to produce a top class strain. It is not usual to be able to obtain winners except by breeding them and developing a strain.**
Winners can be purchased, but this is very much a short term solution and when showing, it is frowned upon for fanciers to buy in just to obtain prizes.

6

INCUBATION
Broody Hen or Incubator?

For small numbers of eggs the broody hen is ideal. If heavy breeds are kept such as Orpingtons, Plymouth Rocks or Wyandottes these will come broody anyway. For bantams a small breed, such as a Silkie or Silkie cross would be used. Unfortunately, the broody may not always be available when required so a small incubator may have to be used.

Once a broody is available, care and attention, especially at the early stages, will pay dividends in terms of overall results. Take time to settle her down and move her to pot eggs in the dark in a private place and make sure she is settled before placing proper eggs under her.

THE INCUBATOR

Incubators are now quite reliable, but still do not give results which are up to the level of a good broody hen. Moreover, she will rear the chicks and they will usually be quite healthy.

Generally it is not worth while to hatch artificially unless there are at least 25 fresh eggs available for hatching, and about 50 is a better number, because an incubator should be reasonably full to operate properly and to be economic. Some chicks will not hatch and a number of the eggs will not be fertile so the number hatching will probably be about 80 per cent of the total placed.

The incubator must be situated in a building or room where the temperature is steady (c. 15.5^0C), otherwise the eggs will be too hot during the day, when the sun is shining, and too cold at night. There must also be the correct ventilation and humidity or the chicks hatched, if they do hatch, will not be healthy.

Do not run away with the idea that using an incubator is simple. It needs experience and practice, and, in between, a number of valuable chicks will be lost.

The size purchased should be sufficient for the likely number of eggs, but not over-size or it will not be easy to run. Remember, it does use electricity which is costly and, on occasions, there may be a power failure, which may be fatal if the period exceeds 24 hours. Be ready for emergency measures, such as placing the eggs in a warm place until the crisis is over.

What is An Incubator ?

An incubator is a machine which has a compartment into which the eggs are placed and which is able to provide the necessary temperature, humidity and ventilation as normally provided by the hen. In addition, since the hen turns her eggs a number of times each day so the minimum requirement is twice per day.

This turning may be done by hand or incorporated into the machine to be turned by operating a lever or by means of a small motor.

Typical *Small* Incubator which allows eggs and chicks to be viewed. Marsh, USA

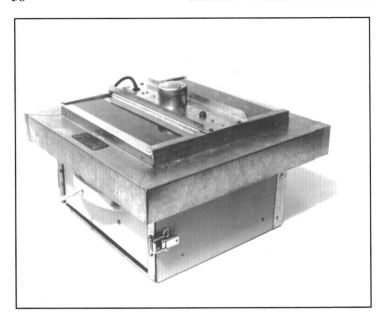

Robust Small Incubator –Still air.

Usually of different sizes and will hold from about 50 eggs up to 200. Select the size for your requirements. Once a larger number of eggs have to be set, a fan operated system becomes essential as well as automatic turning.

For the small poultry keeper the sophisticated system is not essential, but there is a trend to give automatic turning on even the small machines.

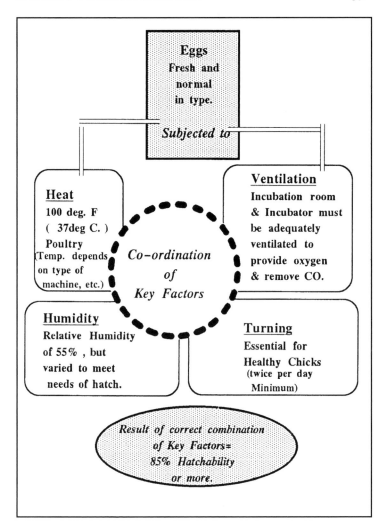

Key Factors for Successful Incubation
From *Artificial Incubation & Rearing*, Joseph Batty

Testing the Eggs for Progress (Candling)

When an incubator is being used it is advisable for a number of reasons, such as contamination from bad eggs and the uneconomic running of an incubator with unfertile eggs, to check the eggs to ensure that the embryo is developing according to the *norm.*

For large fowl the incubation period is 21 days and bantams 19 to 21 days. A temperature of around 37.0^0 C (100^0F) has to be maintained and the instructions with the machine should be followed very carefully. At intervals of 7 - 10 days *and* 14 days it is usual to hold each egg to a strong light and look into the contents and, with experience, to judge which eggs are not likely to hatch.

The Air Space

The air space at the narrow end of the egg provides the *key* to what is going on inside the egg. It varies as the embryo develops so this indicates whether the proper development is taking place.

In addition, as an egg goes through the incubation process, certain changes take place. From the 'germ' the blood vessels develop and these grow into an embryo which at about 19 days has reached a point where it should hatch. The *Candling* indicates the critical situation with each egg and a decision can be made on whether to carry on, or reject the egg by removing it from the incubator.

The inexperienced poultry keeper must tread carefully or many eggs which are fertile will be rejected. When candling, the purpose is to look for life within the egg and life which is progressing, not dead.

The drawings of eggs during incubation, demonstrate the possible stages. At the first candling (7 days) the important stage is the recognition of the blood vessels (diagram) which form a structure which looks like a multi-legged spider, with the body in the middle and the veins all around the yolk. If an egg is broken

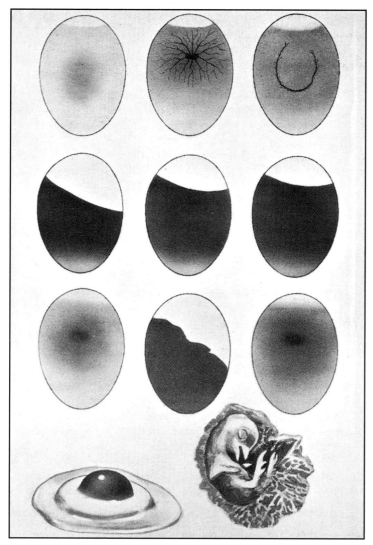

The Egg During Incubation.

Top row (7TH DAY) **left,** clear egg; **centre,** fertile egg; **right,** broken yolk.
Middle row (14TH DAY) **left,** egg dried too much - too little moisture, **centre,** correct drying; **right,** too little drying excessive moisture.
Bottom row left, dead embryo about 5th day; **centre,** ready to hatch 20th day, **right,** embryo died on about 14th day.
Left , at bottom, the egg yolk and small white germinal spot, **right,** an embryo chicken at 14 - 15 th day of incubation.

accidentally or by mistake it will be seen that in the live embryo, the blood is circulating, and even at a few days old, it is a living organism which will develop very rapidly in the egg.

DEALING WITH THE FINAL STAGES

With an incubator there must be great care taken throughout, making sure that adequate water is in the machine and, at the end, from about the nineteenth day, the incubator should be disturbed as little as possible.

It is useful to summarize the rules to follow:

1. Use fresh eggs not more than 7 days old. Mark them with the date in pencil, but use a felt pen to show the date when eggs are put into the incubator.

2. Eggs should be a normal, oval shape with strong even shells (porous shells are not usually hatchable).

3. Locate the incubator in a room which is well insulated and free from sunshine or draughts (fluctuations in room temperatures are better avoided).

4. Keep the water tray reasonably topped up, so that the eggs have adequate moisture. Read the Incubator instructions.

5. Turn the eggs at regular intervals, usually twice per day.

6. When chicks start to hatch try to avoid opening the incubator.

7. Check the thermometer/thermostat at regular intervals so that the incubator keeps steady at around $103°F$ ($39.4°$ C).* Exact temperature depends on type of machine and maker's recommendations.

---

* The exact temperature depends on the machine and therefore the manufaturers instructions must be followed. It is measured at the top of the egg. It may vary between 37.00 and 39.40 C.

Egg Room or Section of Penning Room

This is for the poultry keeper who decides to have quite a number
of birds and hatches as well as selling to a local community, making the hobby quite worth while. For someone keeping a few
birds the storage space would not be so large.

8. Load the machine with a reasonable number of eggs; e.g. at least 25, or so few chicks will be hatched that rearing will be uneconomical.

9. Make sure that eggs are candled at 7 and 14 days. Remove any clears or faulty eggs (See illustration). Be very careful when candling so not to reject live eggs.

10. Have the broody hen ready for mothering the chicks (she must have been sitting at least 14 days or she will not accept them). An infra-red lamp or other method is the alternative.

■■■■■■■■■■■■■■■■■■■■■■■■■■■■

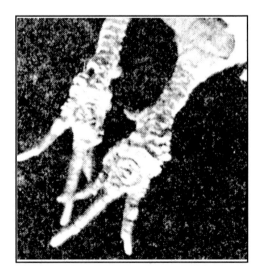

Advanced Stage of Scaly Leg disease.

Regular washing and rubbing with medicinal cream will prevent this problem. Do not breed from birds with this problem or use as broody hens.

BROODIES

Once the chicks have hatched by incubator they can be transferred to a broody hen—one which has been sitting for at least two weeks. Alternatively, a brooder can be used; this may simply be an infra-red lamp placed in a suitable shed with a canopy or metal shade.

In hatching under hens it is *not* wise to use heavy hens of 7 or 8 lbs. weight for *small eggs* such as those from bantams, or you will suffer broken eggs and crushed chicks. If you must use large hens, try medium-weight utility birds such as Buff Rocks, which sit lightly on the eggs and are careful mothers.

Most light breeds, especially the Mediterranean type, do not come broody so a standby has to be available for those who keep these breeds. Also, even the heavier breeds that do come broody should be used on a selective basis, preferring those which are tried and tested or at least from a strain that has a proven record. Pullets (those under a year old) may not have the stamina to go the full distance so try to use a hen of two or three years of age. Avoid those with health problems and do not use a hen with the disease known as 'Scaly Leg' which is caused by a tiny mite which burrows into the scales on the leg, causing disfigurement.

For bantam eggs small broodies are, of course, better and usually the breeder gets enough sitters amongst his own flock; but those who require stocks of broody hens might consider keeping a small flock of Silkie-crosses. These are better than pure Silkies, which have several disadvantages. All with Silkie blood are very tame and take to chicks in an agreeable manner.

They desire to lay and sit again somewhat quickly, whereas many chicks usually require an extended rearing period; they are very susceptible to scaly-leg, which is transmitted to the chicks; and their silky plumage sometimes wraps itself round the necks of chicks and strangles them. Nevertheless, some breeders swear by them and would have no others.

Silkie Broodies Tethered

They must still be watched and returned to the nest after eating and exercising.

Ideal Broody
Silkie crossed with Partridge Wyandotte.

Gives larger size, normal feathers and will to sit.

WORTH ITS WEIGHT IN GOLD
FOR HATCHING & MOTHERING CHICKS

THE CROSS BETWEEN SILKIES AND PARTRIDGE WYANDOTTES IS THE BEST KNOWN. OTHER SILKIE CROSSES ARE GOOD

The Procedures

When a broody hen is seen staying on the nest at night and you require her to be broody, place some pot eggs under her for a few days and place your hand under her when, if broody, she will peck and clutch the eggs with her wings and body. If she flies off in fright then it is likely that she is not yet ready.

Once broodiness is established then transfer her to a broody coop on a pre-made nest of earth lined, basin shaped, with straw and place the eggs berneath her. At times times, getting a broody to settle is difficult because if she is moved she may desert her new nest.

One plan is to have a nest box with lid, which can be moved to a private shed (usually at night time with her on the eggs), and in this way she does not realize that she has been moved. This privacy is essential because, if left in the communal house, other hens will disturb the broody, laying eggs in her nest, so different aged eggs are mixed together.

For normal eggs use a heavy breed, but for bantams a smaller hen such as a cross bred Silkie is ideal.

The broody should be removed once each day for toilet, food and water. If this is not done she may foul the eggs and unless fed daily on mixed corn she cannot sustain the 21 days confinement. Make sure she returns within 15 minutes.

If she is dusted with insect powder before being put to sit there should be no problem with mite, but keep an eye on the posibility by inspecting the vent at 10 days. If necessary dust again, but *not* near the date of hatching because the insect powder could harm the chicks.

Moving the Chicks

Once the chicks start to hatch do not disturb them until dry. Only then should they be placed in suitable quarters. They require no food for 24-36 hours althought the broody hen should he given corn and water when the chicks are moved.

Sitting Box   Dust Bath

**Sitting Boxes, Dust Bath where she can clean herself,
and Water, Grit and Greens.**

The above illustration shows the essentials for the hen. The dust bath, which is inside the shed will be filled with fine, dry earth and fine ash, together with insect powder. She lies down and with a scratching and shuffling motion throws the contents over her, which has the effect of cleaning the pores and preventing lice.

Although Greens are shown, the author believes that they may best be left out of the diet because only hard mixed corn should be fed , with a drink of water. Any soft food or greens may cause looseness of the bowels which may result in soiled eggs. If this occurs the eggs will have to be removed and washed very carefully with warm water and washing up liquid, being careful not to immerse the eggs.

(See opposite).

Typical Nest of Eggs

This shows eggs being cleaned after the hen has fouled her nest. Sometimes this is due to an egg getting broken, possibly because of a thin shell, or it may be that the hen has loose bowels, due to being given soft food or too much water. Wash them very carefully in warm water and washing up liquid and then dry before replacing them on the clean straw which is placed in the nest box.

MAKING USE OF BROODIES

If the intention is to hatch replacement stock then, in the absence
of an incubator, broody hens will have to be used. Unfortunately,
they are not always available at the beginning of a year so, if
needed, they must be encouraged to sit. This may be done by
leaving pot eggs in the nest box and, when a hen is seen on the
nest in the evening, leave her undisturbed and she will thus be
inclined to go broody.

This is the opposite to the end of the season, in the Autumn, when there are often too many broody hens sitting on the
nests. They must then be removed and, if the condition persists,
they should be moved to another shed and run. This discourages
the feeling of being broody because they are now in a strange
place, away from the normal nest box where the eggs were laid
during the season. In breaking off the broodiness some poultry
keepers will deprive a hen of food and water, but this is cruel and
is not really effective. Another approach is to have an isolation
coop with a wire or slatted floor, so the hen cannot get comfortable, which has the effect of cooling down the hen and cutting out
the broodiness. This can be effective, but a change of residence
appears to give better results.

Maximizing Productivity

Many believe that an annual sitting on eggs is beneficial to a hen,
because the limited diet prevents the accumulation of fat. However, sitting too often should be discouraged because the sitting
hen does lose production during the period of being broody.

*If there is no intention of breeding and eggs are the
main requirement the Mediterranean breeds can be kept, but
in a covered enclosure, so they will not fly over the next door
garden !*

In the brown egg breeds much depends on the strain, but Rhode Island Reds, Barnevelders, Marans, or Welsummers will probably be the best choice, dependent upon which is available.

In bantams, any breed will be attractive, but will not have the same laying power as the large fowl. The good layers will produce more than 100 eggs in a season.

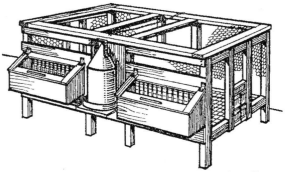

Internal Broody Breaking Coop

These coops are used inside and are complete with water and food hoppers. They have wire floors so the broody cannot get comfortable to sit. A lid would also be on the top. An outside broody breaking coop can be made quite easily in wood, but still with a wire floor and a felt covered roof.

Broody Coop for Breaking Off Broody Hens.

Move into the coop when the hen is not required for sitting in the hope she will lose the desire and start to lay.

7

REARING CHICKS

Once the chicks are hatched they must be reared so that at some stage, probably at around 6 to 8 months they start to lay. There are a number of stages:

1. Chicks which are fed chick crumbs and gradually introduced to broken corn as well.

2. Growers when the amount of protein is reduced by feeding a different ration, including plenty of greenstuff if available.

Fortunately, chick crumbs which contain about 20 per cent protein as well as vitamins are available. Some also include a form of antibiotic against *Coccidiosis* which can be a very difficult disease for young chicks and, therefore, must be avoided if at all possible.

Training the Chicks

The broody hen will watch over the chicks and coax them to take food and water. Generally the naturally hatched chicks are very healthy and strong. She may be kept in a coop with a run attached which can be moved on to fresh ground every 10 days or so. The grass should be kept reasonably short, so they cannot pull up long pieces which would block the crop.

She acts as a trainer, showing them how to eat and drink and warning them of any danger. This is not the case with artificial rearing when some weaker chicks will die because they do not receive encouragement and instructions from the mother.

Later, when feathered, they will be moved into a small shed and run so they can cope by themselves. Different hatches can be put together provided they are of a similar age. Mixed ages may result in pecking and bullying by the stronger chicks.

Feeding on a Regular Basis

Initially it will be necessary to feed four or five times a day, but once they are beginning to be reasonably active and strong, a small trough or hopper can be used for ad lib feeding. If chicks can be allowed in a run in sunshine they come on so much better. However, damp, wet and cold cannot be tolerated so do not expose to inclement conditions.

The water given initially should be in a very shallow dish with a flat stone in the middle, thus giving the chicks an 'island' on which to stand. Unfortunately, in the very early stages they may drown because they exert themselves and then submerge in the water and are unable to get out.

A chick fountain and special food container can be used when they are about 3 weeks old. The food container may have small holes in the top or a bar across which turns so the chicks canniot sit on it.

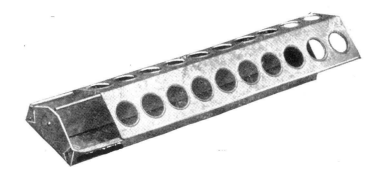

A Food Container for Feeding Chicks from about 1 week old.
From birth it is better to put the crumbs on stout paper or a very shallow dish. For a small number a saucer will suffice. This feeder prevents the chicks getting in the food and spoiling it. Teach them to eat by picking up the crumbs with two figures and simulating a mother hen which would pick up the food and then drop it to show chicks what to do. Once one starts the others imitate.

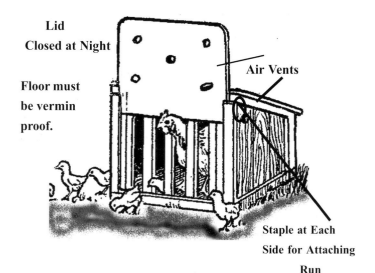

Lid
Closed at Night

Air Vents

Floor must
be vermin
proof.

Staple at Each
Side for Attaching
Run

Chick Coop Ideal for Fine Weather

The author prefers indoor chick boxes with a wire netting front

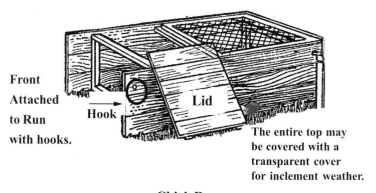

Front
Attached
to Run
with hooks.

Hook

Lid

The entire top may
be covered with a
transparent cover
for inclement weather.

Chick Run

This is attached to the coop at the front and should always be in
position when the chicks are not being observed. Putting hooks
on the corners to secure to the coop is advisable because otherwise a fox or other animal may push the run over or to one side.
A wire netting floor is advised because this prevents any predators tunnelling under.

WEANING

Chicks with a mother hen soon become independent and want to move away seeking food. When they are feathered at about 8 weeks of age she will want them to be on their own and, once she gets the urge to lay again, will push them away, even pecking them. This is all part of nature's way of making the chicks grow up.

At this stage the chicks should be placed in a separate shed with a run. A small place will suffice, but with some internal floor space so they can be under cover in bad weather.

Chicks can be transferred on to Growers' mash or granules from about 3 weeks of age. Introduce this gradually, but keep feeding the broken corn and the chick crumbs.

PURCHASING DAY-OLD CHICKS.

Replacement stock should be obtained every two years, when the old birds are losing their ability to lay an acceptable number of eggs. With a commercial farmer the hens will be discarded after one year, two years at the most. However, done as a hobby or part-hobby it means that birds that are still laying a reasonable number of eggs, say, around 100 per annum, could be kept on until about 3 or 4 years of age. However, any non-productive birds should be culled and, if plump, they can be killed and eaten.

If you buy the chicks from outside they must be reared by letting a broody sit on pot eggs until the day-old chicks arrive, and then remove the eggs and put the chicks under her after dark, when she will usually brood them as though they were her own chicks. Try with the strongest chicks at first, but watch carefully, because some surrogate mothers will refuse to take the strange chicks and may even kill them.

The alternative approach is to rear by using an infra-red lamp or other method of rearing artificially, covered later.

A Few Don'ts.

Don't "set" stale eggs. Put the eggs you intend to hatch under the broody as soon as possible after they are laid—the fresher the egg, the stronger the chick is likely to be.

Don't set a broody in the ordinary poultry house, but once quite ready, put her special nest-box away from the other hens.

Summary of Rules to Follow

1. Do not feed the sitting hen at odd times; see that you feed her at the same time each day—if not, the hen gets restless.

2. Do not give the sitting hens soft food ; feed on maize only or hard, mixed corn, because this grain contains more oil than wheat and oats, so it helps to keep up the body heat. As noted, an alternative of mixed corn can be given.

3. Do not replace infertile eggs with fresh eggs when you remove them on the seventh day , because they will not hatch at the same time. Moreover, eggs which are hatching will need different conditions from those which are incubating in the normal way.

SEXING THE CHICKS

At the time of weaning and possibly *earlier* with the light breeds it should be possible to sex the chicks, thus dividing them into *pullets* and cockerels. However, any decision on which are the best birds may better be left until the growers are fully feathered, because often the early chicks do not mature as indicated in the early stages.

Above: Heavy Breed; eg, Sussex, Cockerel on left. Usually possible to be certain at 9/10 weeks.

Bottom: Light Breed Cockerels with Broody (Heavy breed). Note embryo spur.

Sexing the Growers

ARTIFICIAL REARING

Rearing chicks requires great care and attention, especially during the first week. Unlike many other young birds, chicks are quite independent and can peck for food from about 2 days' old. They need warmth (90 degrees reduced by 5 degrees each week until around 65 is reached; 35^0 C to 21^0 C), chick crumbs and water in a small, chick fount. Broken corn and chick weed can be given as well.

There are many possible systems to use for rearing chicks and each has its merits, but where only a few are involved an infra red lamp with a dull emitter is the best idea. The lamp is lowered to just above the floor and tested to make sure it is the correct temperature. Following that it will be raised a centimetre each week or other period to reduce the heat so that after a few weeks the chicks will have heat only at night and then, once feathered, they will not require the lamp at all.

A bright light may damage the eyes of the chicks so the dull heater is preferred. However, watch the chicks carefully because if they huddle together the temperature is too low and if they lie with their necks outstretched on the ground, gasping for breath, it means the heater is too high. If allowed to get cold in the

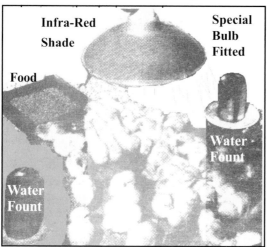

Infra-Red
Shade

Special
Bulb
Fitted

Food

Water
Fount

Water
Fount

Infra-Red
Lamp.
Raised to
reduce heat.

The heater level
starts at 45cm
from floor and
then raised until
chicks
hardened.

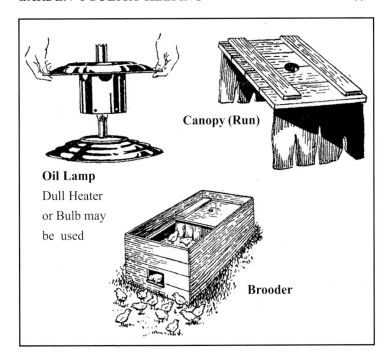

Oil Lamp
Dull Heater
or Bulb may
be used

Canopy (Run)

Brooder

Home made Brooder and Canopy Run

The canopy fits over the brooder so the chicks can run out when strong enough. This hardens them until feathered so that at about 8 weeks they can ***run on***, but much depends on the climate and time of year. Spring hatched chicks thrive best because they can depend on some sunshine and the days get longer.

RUNNING ON

Those reared artificially will be ready to go without heat at about 8 weeks and should be placed in a small shed with run so they can grow in the sunshine. A lining of hay may be helpful to keep the chicks warm until fully hardened. In very cold, frosty weather keep them indoors. A rearing pen or a large shed will be ideal.

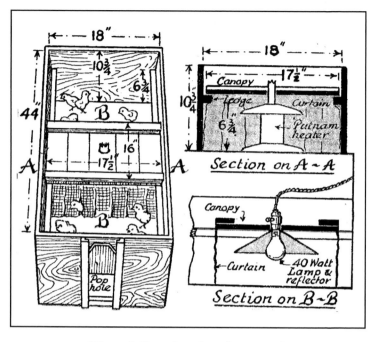

Plan of Brooder *(previous page)*

When an electric lamp or heater can be used this is more reliable. The bulb should be the coloured type so the chicks are not affected by the strong light; usually red or yellow bulbs can be purchased.

Make sure the brooder is achieving the correct temperature before placing the chicks inside. The safe way is to have it running for a few days before chicks are due to hatch.

Separation of cockerels is essential because the unwanted birds may be sold or fattened for the table. With show birds it will be necessary to keep them longer until the potential is realized.

The signs are as follows (also see page 77):

1. Light Breeds

(a) Cockerels

Comb showing, face reddening, long tail, cheeky appearance, long neck and rather leggy.

(b) Pullets

Comb small and yellow colour, longer body, short tail, lacking confidence of *cockerels.*

2. Heavy Breeds

(a) *Cockerels*

Coarse heads, comb showing, feathering *on* back slow, thick set with strong legs, upright carriage, greedy and aggressive.

(b) Pullets

Smaller head, no comb growth, body well feathered, tail well developed, shorter legs, not agressive.

Disposing of Cockerels

The light breeds are of limited value for food so try to sell to other small poultry keepers. The heavy breeds can be killed at 3 months when they should be about 3lb. (1.50k.). Specialist sales are held each year for rare breeds and are held in many places, but the Rare Breeds Sales at Stoneleigh and Salisbury are very well known. Many bargains can be acquired, although some of the scarce breeds of top quality do reach quite high prices. Once top quality stock is obtained the resulting offspring can also sell at very good prices.

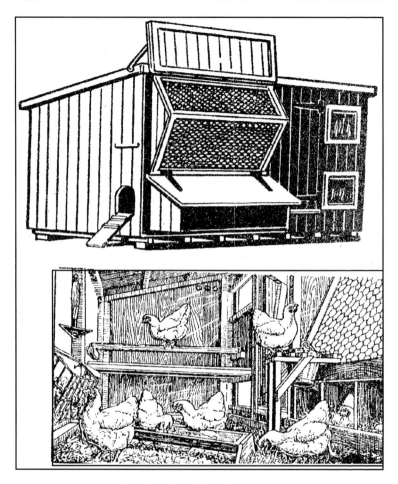

Poultry Shed, Semi-intensive.

This may be used in all seasons because the birds have scratching room in inclement weather.

This is a very clever design because it allows the birds plenty of exercise.

Note the Shutters which can be closed in winter. Should be covered in perspex, glass or plastic so the light is still present.

The bottom illustration shows the interior. It is a useful exercise to see how many essential items you can spot.

8

MANAGEMENT

Daily Routine

If poultry keeping is to be successful a daily routine must be established. This is primarily to feed and check the water, but it should also include a check on the birds to make sure nothing is wrong. The feeding of greenstuff and litter from the garden should also be done,.

Eggs have to be collected whilst still fresh and put into trays. If necessary they should be washed or cleaned dry with wire wool.

Feeding

There are two schools of thought on feeding summed up as follows:

1. Ad Lib Feeding

The birds are supplied with a hopper which allows the mash or pellets to run down into the base from which the birds feed. Greens can also be fed or the birds run on to grass, but for the main feed there is total reliance on the pellets or mash. Pellets are more palatable and cleaner, not clogging up the water from the birds drinking.

2. Hand Feeding Two or Three Times Daily

Pellets may be fed in the morning followed by scraps and greens at mid-day and then mixed corn in the early evening.

It is important not to overfeed for any surplus will encourage rats or mice and hens which are fed too much are in any case slothful and, once too fat, will lay very few eggs.

For the fancier this is the best approach, but when it is not possible (due to business commitments) a compromise will be necessary, such as having pellets in a hopper and once daily throwing a handful of corn per bird early in the morning or in the evening.

Special Feeding Requirements

When exhibiting birds it should be appreciated that the hard feathered varieties such as Old English Game must have mixed corn to retain the tight, glossy coat of feathers. If fed on pellets alone the birds become too feathery.

This applies to Old English Game, Indian Game, Asil, Malays, Modern Game, Rumpless Game, Shamo, and Tuzo. It should also be applied to birds which, although not Game birds, do have quite tight feathering such as Sebrights, Sumatra Game and Rumpless Game, and excessive soft food would make the plumage too fluffy underneath. The same rule applies for large fowl and bantams.

However, with any breed which is fed on mixed corn alone no records will be broken for egg laying, simply because the necessary protein and other requirements are not included in the food. In any case, the hard or tight feathered birds are not bred for laying so the potential is not present anyway. But, with a hobby, a record number of eggs are not the most important. Old English Game are really a compromise, laying a satisfactory number of eggs as well as being a good table bird.

Malay Bantams
Not good layers,
but full of character
and most interesting
to keep.
Hard feathered so
mixed corn should
be fed, except when
breeding.

Watering

A regular check on the water will be essential. If time permits a daily change of the water will be necessary, especially when the weather is very hot. In winter when the water is frozen it will be essential to empty the water fount in the evening and then, using a watering can, replenish with sufficient water for the day.

If there is snow the birds outdoors will peck it to obtain the 'drink', but this should not be over done because fresh water is better., Sometimes, in very cold weather, the birds seem to *appreciate* warm water. If the founts are really frozen hot water can be poured in the base so the water is freed.

If there is snow the birds might be better kept indoors in a scratching shed. The dedicated fancier *will* ensure that suitable accommodation is available for this purpose.

There is no standard amount of water per bird per day, but to be on the safe side allow 2 pint or litres per adult bird and make sure they always have plenty. If dry mash (powdered food) is fed, more water will be taken.

Wet Mash

The feeding of a wet, crumbly mash can stimulate the appetite of birds and this was a common practice, but, nowadays, the ad lib feeding which is labour saving tends to be used. If a person has the time mixing a wet mash can be quite beneficial to the birds, especially in cold weather. Bread, scraps of cheese and meat, and other tit bits can be added, all waste from the household. However, it muse be realized that the ration must be balanced and contain enough protein to sustain the body and egg laying,

In protein terms, we must think of the percentages as the following:

1. **Chicks - about 20 per cent.**

2. **Growers - 14 per cent.**

3. **Layers - 16 to 18 per cent.**

Fish and meat meal is more effective *than* grotein from peas or soya. Feeding more protein than stated is wasteful and if fed to Growers may bring the birds into lay too quickly which will reduce the size of the egg.

The premixed food from the feed mill or poultry food merchant shows the level of protein, this being required by law. Corn has a protein. level of around 10 per cent so it will be apparent that a complete diet of corn is not advisable, but it does delay maturity.

Purhasing the Feedstuff

If six or eight layers are to be kept they will eat about 25 kilos in 3 weeks and this should be purchased in bulk in 20 or 25 k. bags. Buying small quantities is not economic.

Once an initial stock is acquired, in the interests of economy, a bag of mixed corn may be purchased every 3 weeks and the layers pellets 3 weeks later, thus budgeting for the cost. Make sure there is maize in the corn, because wheat alone is inadequate. Do not over stock because the pellets may not keep, especially in damp weather.

Breeders' Rations

When birds are in the Breeding Pen, say, a cock and three hens, they should be fed with a high protein food. Breeders' or Layers' pellets should be given in a hopper for ad lib feeding and the corn linited. This food ensures that adequate protein is present which helps fertility and avoids producing weak chicks.

The ideal is to have the birds on a lawn or in an orchard where they have plenty of grass because this keeps them fit and helps the fertility.

Grit

There should be a regular check made on grit available, both flint and limestone. Without grit the birds cannot function properly.

At the first sign of soft shelled eggs or shells which are uneven take positive action. The problem may be a shortage of grit or the hens are too fat or they are not taking the grit. Soft shelled eggs can lead to **Egg Eating** which can be a curse in the poultry yard, and very difficult to eliminate once the habit starts. Having the nest box in darkness by having a curtain across can prevent the eggs being seen. Filling an egg with mustard, or having a nest box which slopes so the egg rolls away, can also be successful, but the answer is to prevent the problem starting in the first place.

In the pellets purchased from the mill grit is included, but some hens may need more and do *not* pick it up. In these circumstances sprinke fine grit in a wet mash until the eggs are well shelled. If the problem persists try to spot the hen and then feed her grit by hand – a little fed into her beak will suffice. Once she is getting enough grit the problem should disappear.

Soft shelled eggs are a menace and must be eliminated.They are of no value and can lead to egg eating, which is difficult to stamp out the problem.

Storage

As noted earlier, there should be adequate provision for storage. A metal hopper or even a dustbin will suffice to keep the food fresh and away from mice.

A separate Store Room combined with a Penning Room to train birds, or carry out other functions connected with the birds, will be found to be a great help.

Besides the normal equipment and sundries required it will be essential to have a medical chest to deal with injuries or ailments. The requirements are listed on the diagram showing the layout of the Penning/Store Room and reference should be made to that illustration.

Weekly Routine

There are many tasks which will have to be done at the *weekend* or an a free day. These include the following:

1. Renew Litter & Nest Box Material

The floor of the poultry house should be strewn with litter to a reasonable depth. This should be checked each week and any soiled material removed and replaced with new.

The best material is dry peat moss or wood shavings. The latter is cheaper, but does not mix so well with the droppings to make manure.

Straw and hay may also be used, but they do not absorb the moisture to anywhere near the same extent. These can be used for nest boxes.

2. Clean Droppings Board

If a droppings board is placed beneath the perches this should be cleaned by removing the faeces and sprinkling sawdust, peat moss, sand or dry earth. Remember this is excellent manure for the garden, but it is also quite strong and is best allowed to dry under cover before using.

A scraper may be found quite useful for ensuring all the droppings are removed easily.

3. Pick Up Birds and Handle Them.

This might be best done in the evenings when the birds have gone to roost. If they have to be caught up use a large net and, if they escape, do not chase them too much because this will induce alarm and may cause them to stop laying; open a pen door and *gently* drive them in.

A Collection of Breeds

All offer something different: Leghorn, prolific white egg layer;
Black Orpington, Brown eggs and suitable as table bird; Wyan-
dottes, good dual purpose breed; Langshans deep brown eggs, but
Modern Langshan mainly a show bird which is quite tall.

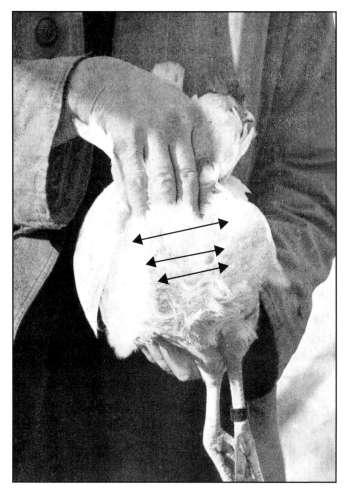

Testing for Laying Hens

The productive hen will have a wide gap between the pelvic bones. Some call this the "Three finger Test" and this measure is a rough guide using an average person. The important principle is that the bones should be wide enough for eggs to pass through and the vent itself is raised, soft and pliable. If the pelvic bones are close together, with no gap, then the pullet has not started to lay or the hen has ceased laying.

Bantams have a smaller gap, but the principle is the same.

WATCHING FOR HEALTH PROBLEMS

Health and condition can be seen from handling. The body should be firm, but not fat. If the vent is *inspected* it will be found to be large and moist. If two (possibly three) fingers can be placed between the pelvic bones the hen is usually laying. However, if the space is very narrow, it means no eggs are being laid and if this persists the hen will have to be culled; ie, disposed of.

At this inspection, check for lice at the vent and dust with insect powder. If the feathers at the vent become soiled remove therm and rub the area with a light coating of sulphur ointment which kills any lice. If there are *cluster*s of tiny eggs pull these off before applying the ointment. If left, the lice hatch out they thrive on the accumulated droppings on the feathers and make the vent sore. Accordingly, it is vital that an inspection is made on a regular basis; in warm weather once every ten days is advised and vaseline rubbed on the area (sulphur ointment is ideal if it can be obtained) and this prevents any problems

Sometimes the feathers on the back become very worn and broken due to an over enthusiastic cockerel. In this case he may need more hens or he should have two pens of birds, letting him to each on alternate days. Again, treat the patches with sulphur ointment. Feather mite may occur when the feathers are eaten away and, once noticed, the birds should be caught up and washed with warm water, including a mild disinfectant.

If a hen becomes damaged through any cause she should have the area bathed and a medical disinfectant added, such as **Dettol.** In the case of a bad injury or she is being pecked, separate the hen from the main pen until she recovers. A small coop and run is adequate for this purpose.

If on inspection a hen is permanently on the nest, then, if not required as a broody hen, move her to a special pen with a wire or slatted floor which will break her of the habit. Alternatively place her in a new *run* with a different cock and this may break the broodiness and she will return to lay.

Simple Nest Boxes

Litter should be renewed weekly. Hessian curtains strung across the front, with openings for access, should give privacy in sem-darkness.

**Scraper for removing any Stubborn Droppings
from the Droppings Board.**

PERIODIC TASKS

The runs should be kept in good repair and when foxes are in the area make sure the birds are locked up each evening. Wire should be sunk into the ground so that no tunnelling is possible. This should be checked at regular intervals.

Door hinges and locks should be oiled and the sheds should be painted or creosoted about once a year. If mite is to be discouraged the perches and droppings boards should be removed and all crevices should be painted with paraffin or creosote.

This spring cleaning can be done before the breeding pens are made up at the beginning of the year. Allow the inside to dry before putting birds back because creosote does burn and affects their breathing.

Culling Birds

Sadly we cannot became too attached to our feathered friends because if they are to be profitable they must be culled at the end of the laying season and replaced with new stock.

This can be done in July or August and will be based on whether they 'have paid for themselves' in terms of egg production. There is no exact figure for a *standard,* because white egg layers tend to lay more and the heavy breeds lay smaller numbers than light breeds, but they have plump bodies for eating.

However, there will be a leader in the hens kept and her total will be a guide; about 125 eggs is probably the minimum for normal breeds, as opposed to, say, ornamental bantams, which may lay about 50 eggs in a full year.

This can be done by inspection and observation throughout the year. If a thorough job is to be done the process of "trap nesting" will be necessary, whereby birds go on to a nest to lay and by the use of a special door are kept on the nest and then removed and a record kept of the hen and the egg laid. This is not always a practical approach for the amateur poultry keeper.

LIKELY CULLS
Those to weed out.

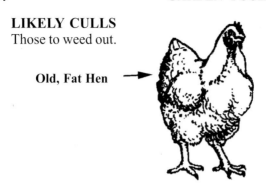

Old, Fat Hen →

Early Moulter & Poor Layer

Gasping, Ashmatic Bird

←Under sized Hen
Lays Small eggs

Perpetual Broody
Rarely Lays →

Drooping Abdomen
Constant down behind
indicating too fat or prolapse.

Checking the Performance of Each Hen

The alternative method is to check each hen and then, taking into account the history of the bird, make a decision on whether to keep her. Factors to consider are as follows:

(a) **How many times broody;**

(b) **Condition throughout the year;**

(c) **Whether laying at the time of inspection;**

(d) **On inspection, whether the signs indicate a winter layer – usually if she has finished laying in July or August there will be little hope of continued production;**

(e) **General health and condition;**

(f) **Early and lengthy moulter.**

Checking a Production Record should make it possible to spot the likely 'winners'. Those showing negative qualities should be culled.

Moulting

When the Autumn comes the birds will lose their feathers and will be out of condition. The good layer usually moults quickly and recommences laying. ***Very exceptional birds, given food with plenty of protein, may lay for most of the period of the moult. Also give linseed, cod liver oil, and high protein foods.***

The poor producers moult early and take many weeks to come back into lay. Some may not lay at all until the Spring. Such wasters have to be weeded out and may be sold, but without any guarantee of quality laying, or killed and eaten. Seriously diseased birds should be killed and incinerated.

The moult follows a definite sequence and therefore the poultry keeper can see what stage has been reached. This has been researched by practical students of poultry, and the details are described and illustrated on the following pages. Obviously, in practice, the exact way feathers are being replaced, is difficult to follow, but a general idea is useful.

MOULTING BY NUMBERS

The order in which the feathers moult follows a definite pattern in the following sequence:

1. **Head feathers drop , followed by the back of the neck.**
2. **Back, fluff, main body feathers and first primaries at about the same time.**
3. **Tail.**
4. **Secondaries in wings start to drop.**

If a bird stops laying before moulting, the first primaries drop only a few days before the secondaries. But if a bird lays on after starting the moult, there is usually a delay, sometimes of several weeks, before the secondaries drop.

It is rare for a bird to lay after the dropping of the first secondaries, and this is a good guide to the end of production. When the first secondaries fall, laying is usually at an end until after the moult.

Tail moult does not follow the same pattern. Occasionally the tail moult is scattered throughout the tail, and in a good layer it is not unusual to find the entire tail has moulted at about the same time.

In the wing the primaries are moulted in a definite order. Each wing has ten primaries, occasionally, when the wing is spread out, is a small feather, the "axial ".

The first primary to drop is the one next to the axial feather, and in a normal moult the next is dropped two weeks later, and so on, in order, up to ten, at two-week intervals.

The new quill starts to grow as soon as the old feather is out and it requires six weeks for one to grow to full size.   It will be seen that when the fourth feather is dropped, No. I will be fully grown.

Assuming that the hen stopped laying when the first primary was dropped, by allowing six weeks for the first feather to be fully grown, and two weeks for each additional feather, we can get a good idea when a hen stopped laying. It will be seen that a normal moult would take 24 weeks for all ten primaries to reach full size, This does not occur often, except with poor layers. The good layer will often moult two or more feathers at one time, and therefore takes a much shorter time to moult.

In such cases, all feathers dropped at one time should he counted as one feather when estimating the length of time a bird has been in the moult. If a hen drops out of lay during the summer she may moult one or more primaries, then stop moulting and come into production again. This is known as a " vacation " moult. When she starts her full moult later, she will drop the next feather in sequence and moult in order of the remaining primaries. She then goes back to the one next to the axial and moults again those which had been re-newed in her vacation moult.

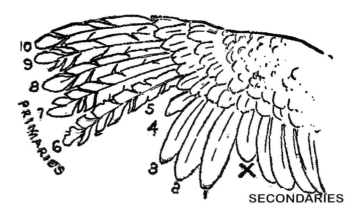

NOTES: Primary No. 1 next to Axial Feather X, drops at start of moult. Numbers 2,3,& 4, etc, drop in sequence at intervals of 8 to 14 days. The primaries are renewed in the same order. Thus 1, 2, 3 and 4 (see sketch) are new feathers. Number 5 has dropped and 6, 7, 8, and 9 are old primaries.

CULLING: A hen such as that described in August should be culled and the same would possibly apply in September to November. However, the late moulters usually make a quicker moult than the early ones. The decision would depend on the overall position.

MOULTING HEN which should be culled if like this in July and August.

Feathers are dropped in the regular sequence of the numbers 1 to 9, but are not renewed in the same order.

Source: Details of Moulting Sequence from *Keeping Chickens for Profit*, Alan Thompson, now out of print.

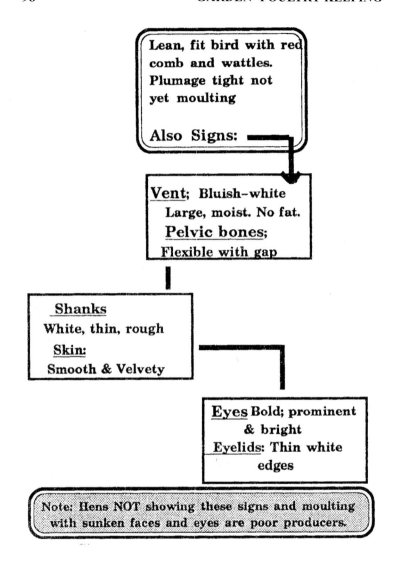

Lean, fit bird with red
comb and wattles.
Plumage tight not
yet moulting

Also Signs:

Vent; Bluish–white
Large, moist. No fat.
Pelvic bones;
Flexible with gap

Shanks
White, thin, rough
Skin:
Smooth & Velvety

Eyes Bold; prominent
& bright
Eyelids: Thin white
edges

Note: Hens NOT showing these signs and moulting
with sunken faces and eyes are poor producers.

Signs of A Good Layer

May be an excellent winter layer which is quite important

culled.

Lighting

If possible a small wattage bulb should be used to give the birds light (to bring up to a total of 12 hours per day). Usually production is increased by around 20 per cent by giving theextra light so the exercise is very worthwhile.

A time switch can be used or the light can be switched on in the evening and then turned off before going to bed.

A Basic Lighting System

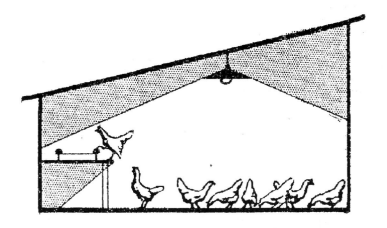

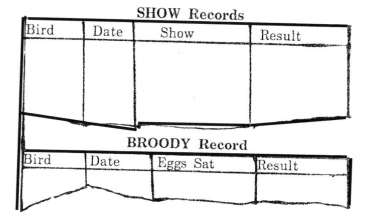

EGGS LAID						
	Pen No					
Bird No (or name)	1	2	3	4	5	Remarks
Date						

SHOW Records

Bird	Date	Show	Result

BROODY Record

Bird	Date	Eggs Sat	Result

Top: Basic Production Record which can be ruled on to a stiff card and then kept near the nest boxes.

Middle: Details of show entries and results.

Bottom: Broody record of each hen and the results obtained.

KEEPING BANTAMS

Miniature Fowl Delightful and Profitable

The small, pygmy-like poultry are known as **bantams,** which are about 25 per cent of the size of the equivalent large fowl or are around 20 oz. (550g) to over one kilo, if they are natural or true bantams. They tend to be quite diminutive, the larger weight being the hard feathered bantam, the Tuzo. The remainder all tend to be at the lower end of the scale.

True Bantams

These are bantams where there is no large equivalent so the assumption made is that they always existed in the small size. They include:

1. Belgian Bantams   2. Booted Bantams

3. Dutch Bantams   4. Japanese Bantams

5. Nankin Bantams   6. Pekin Bantams

7. Rosecomb Bantams   8 . Sebright Bantams

9. Tuzo Bantams.

They are ornamental breeds, kept for pleasure and for showing so do not expect top laying from any of these. However, they are quite delightful, often beautiful and become quite tame.

Some of these such as the Japanese and Pekins can be kept in quite small sheds with an indoor run, with shavings on the floor, so they do not become dishevelled with mud or water. They are most suitable for children or the disabled because they become quite tame and would not generally fly away.

Some are of great antiquity and it may be that they existed before the large fowl, because they are around the same size as Jungle Fowl, the ancestors of domestic poultry.

* **For a full history see *Bantams & Small Poultry*, Joseph Batty, BPH.**

Utility Bantams

Obviously the bantams which derive from the large breeds are more likely to be better in terms of laying and some, like Indian (Cornish) Game, are good table birds so any culls can be utilized for the table. A fully mature Indian Game cock can reach 2 kilos (4.4 lbs) or more so there is considerable meat on such a bird.

Generally speaking, the better bantam layers produce about 125 eggs per year, mainly in the Spring and Summer months. This is half the number of the *best* large breeds, but the eggs are a reasonable size and they only eat about half as much as a large fowl. Some lay brown eggs, whereas others lay white or tinted. The colour of the egg in the large breed is not always achieved in bantams, but in recent years there has been improvement on this aspect. For example, *large* Barnevelders lay very dark brown eggs, but for a long time the bantam eggs were a much lighter colour.

Because of their small size bantams are excellent for the garden because:

1. Smaller houses and runs.

2. Less food consumed.

3. Good egg size (35 - 40g.) in relation to food eaten.

4. Become quite tame.

5. Easy to handle, even by children and the disabled.

6. Easily transported to shows; the special baskets can be put in the boot of a car (must be checked to make sure not too hot and properly ventilated).

7. If a cock kept not as loud a crow.

8. Good demand for popular breeds so any spare stock can be sold to cover food costs.

Bantam Shed & Run Combined

This shed is off the ground which means predators cannot dig underneath. It has shutters for closing in inclement weather. Note the overhanging roof which keep the rain out, yet allowing plenty of fresh air through the wire. In summer the lower shutters can also be removed.

Internal Dust Bath for the birds to cover themselves with fine earth and insect powder thus keeping clean.

Bird in box scatters the earth and throws it across her, at the same time turning so she gets the earth on all parts of her body.

NOTED EGG LAYERS

The *breed* denotes the type of bird in terms of size, shape and general characteristics, and this is sub-divided into *varieties*, usually the different colours. In addition, there is the *strain* of the breed and some strains are more productive than others, because the owners have selected for laying, making a note of the number of eggs laid by each hen. If there are more than three hens the breeder must use a *trap-nest system*, which allows the hen to be trapped in the nest until released and the egg laid is noted. Usually the hens are rung with a numbered ring, thus making the process quite simple.

Unless the poultry keeper has plenty of time on hand, the trap-nest system cannot be used, because it does mean that hens have to be let out on a regular basis, soon after they have laid. A record card can be kept near the nest boxes.

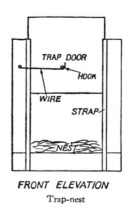

FRONT ELEVATION
Trap-nest

Trap Nest Front

This would fit on to the front of a nest box and the hen pushes into the interior and after she lays is removed and released.

White Leghorn

White Pekin

Millefleur Belgian

White Wyandotte

Spangled Old English Game

Rhode Island Red

Selection of Bantam Hens

Belgian and Pekins are true bantams of the ornamental type. The remainder are commendable layers.

CLASSIFY THE BANTAMS

Bantams can be classified into categories so a selection can be made, thus:

Layers of White Eggs

Anconas, Andalusians, Minorcas , Leghorns (many colours), Hamburghs.

Layers of Brown Eggs

Barnevelders, Welsummers, Marans and Langshans.

Table Birds

Indian Game, Ixworth, Sussex, Wyandottes, Old English Game and different crosses of these. As noted, Indian Game cocks can reach 2.2 kilos.

Dual Purpose Layers

Rhode Island Reds, Orpingtons, Sussex (Light, White, Buff and Speckled), Wyandottes (White, Black, Partridge, Columbian and Laced) and Australorps.

Specialist Breeds

Old English Game bantams have a very large following and exist in many attractive colours. Sebrights and Rosecombs, both true bantams, are extremely attractive, the Sebrights being laced in black on a gold or silver colour, whereas the Rosecombs have very large combs and are small and dainty.

Other breeds which are extremely attractive are Frizzles (feathers growing the wrong way), Polish with large crests, Modern Game, tall, elegant bantams in well defined colours, and Scots Greys with barring which is quite attractive.

There are many other breeds which can be selected so that the fancier can keep and breed from very attractive stock, which will give great pleasure.

Minorca Columbian Wyandotte

Indian Game Partridge Wyandotte

Barred Rock Rhode Island Red

Variety of Bantam Hens

FEEDING

Feeding is along the same lines as large poultry. The *small* type of pellets can be fed, augmented by mixed corn, which is essential for keeping hard feathered birds, such as OEG and Indian Game, tightly feathered.

Give them plenty of greenstuff from the garden in the form of grass clippings, chick weed and leaves, which they will peck and eat. This is vital, especially when they are kept indoors. Household scraps can also be fed, but never any mouldy bread or other food which is deteriorating.

Feeding Chicks

Chicks are fed chick crumbs, and water is given on a saucer or shallow container. Milk can also be given which is quite nourishing. Never leave any stale food or water in the pen because this may cause serious indigestion problems, and loose bowels, with a sticky "back end", which must be dealt with because this could cause the chick to die from the problem that exists internally.

For the first 10 days watch the chicks very carefully. Make sure the hen does not scratch shavings into the drink and food containers, with the result the chicks starve.

Many breeders rear indoors in wooden boxes with wire fronts; the type used for breeding cage birds can be quite useful. The chicks are then protected from predators and from any disease which might be picked up if running outside. Loose bowels and the complaint *Coccidiosis*, which is a disease of the bowels, are easily contracted from the ground, from bird droppings. It is also aggravated if the chicks run on the grass too early, especially in wet weather.

Special care in the first few weeks will pay dividends so the chicks are allowed to grow sturdy before being exposed to dangers. Once they are strong they can be moved out or into a large shed. However, do not mix chicks of different ages.

10

SOME PRACTICAL CONSIDERATIONS

Purchasing Birds

The decision on the type of bird to purchase should be made after considering the relevant factors in earlier chapters. This will be influenced by such matters as:

1. Space available; if limited to a small garden then a trio of bantams should be the decision;

2. Whether layers or dual purpose birds are to be kept, eg, Leghorns or Light Sussex;

3. Is the purpose to supply fresh eggs and table birds for the household;

4. Is any attempt to be made to sell products, eg, eggs and roasting chickens;

5. Are birds to be exhibited and if so which breeds are likely to offer the worthwhile challenge; eg, an exotic breed like Modern Game or a practical breed like Welsummers or Rhode Island Reds;

6. How much time is available for fitting in the work required;

7. Are you prepared to handle birds and, if necessary, kill them for eating or culling: if not someone else will have to be called on.

Simple Open Shed – Ideal for rearing

A door would be essential for protecting the birds, but useful
when in a paddock as well as a permanent shed.

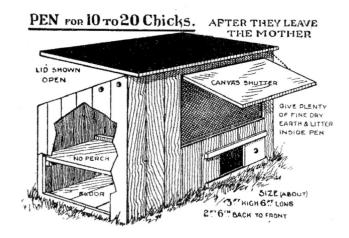

Chick Rearing Shed

After the chicks are weaned from the mother hen they should be
placed in a separate shed with a run attached.

Possible Sources of Birds

Birds may be purchased through the specialized press such as *Fancy Fowl* or *Cage & Aviary Birds.* Alternatively, they may be found locally by scanning the local papers under 'Pets' or 'Livest;ock'.

Often bargains can be obtained, but do not buy very old hens because they will lay very few eggs. Remember, the useful life is about two years so unless you are buying high class stock intending to breed from them do not be tempted with old stock. Stock turned out of battery cages are unsuitable and generally look rather raggy and unattractive, so do not be tempted with these birds.

The hints given earlier on healthy stock should be applied and the sections on Culling should be studied. Generally the heavy breeds will lay around 150 - 200 eggs per annum and the light breeds will produce about *250*, but will have very little meat on them for eating. The heavy breeds can be used for hatching eggs, but the light breeds do not; come broody.

Some breeds are classified as 'Rare Breeds' because they have been forgotten by poultry farmers who have concentrated on commercial layers. For those who wish to keep rare breeds the **Poultry Club** or **Rare Poultry Breeds Society** should be contacted. For advice on how to keep breeds contact **The World Bantam & Poultry Society,** available through the publishers.

Select a breed which appeals to you, but be prepared to supply the appropriate housing because some, such as Yokohamas (long tailed Japanese) require facilities which will not damage the tail and Polish will need water founts which avoid wetting the crests. Some of these exotic breeds are quite reasonable layers and very attractive, but do need that special effort to find stock and then give them the attention they need, trying to bring the breed back to a high standard so that it is not lost. Generally, they are no more expensive to keep and the effort is very worthwhile.

IMPACT OF THE ENVIRONMENT

The 'fancy breeds' such as Brahmas or Cochins should have a covered run and be kept out of the mud or the foot feathers will be damaged. Bantams cannot stand the bad weather as well as tile utility large fowl and must have a cosy house and run for the Winter months.

Some breeds, mainly the heavy types, can stand heavy soil, but they cannot stand stagnant water lying around and mud, which will occur in runs which are not managed properly. Sussex are quite hardy and one breed, the Marsh Daisy, was bred to thrive in difficult conditions.

Unfortunately, following more than 50 years of laying cages for which special hybrid breeds have been developed, it may be found that the only easily obtainable stock are these birds.

Except for the cruelty aspect, these hybrids were excellent for the purpose for which they were intended; namely, close confinement and maximum production under controlled conditions. However, they have not been developed for free range and are not fully tested for that purpose. The author has visited so called 'free range' farms where these birds are kept and it was found that they were reluctant to go outside the large shed in which they were kept.

On the other hand, the standard breeds are hardy and have been accustomed to foraging for food and eating the fresh grass and natural food which is found in abundance when fowls are kept outside. Even when there is limited space in a garden, lawn clippings and weeds provide excellent fresh food which birds enjoy so much. Throwing weeds into the run with soil attached will build up the run so that it keeps fresher than leaving it in its original state. This waste is converted into food by the hen.

Therefore try to obtain sound stock with a good pedigree for laying and, when required, for table purposes. Surplus cockerels can be fattened at little extra cost and provide a valuable source of food.

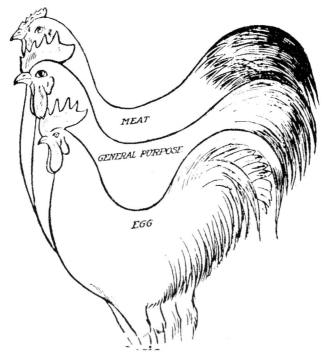

Basic Types of Fowl

This approach distinguishes by main purpose of the breed or by characteristics. Laying types are developed well at the abdomen (back end) whereas the heavy, meat breeds have full breasts.

Leghorn Hen
Excellent Layer. Well
developed back end.

Jubilee Indian Game Cock
Broad breasted for meat.

Polands (Polish)

These are rather exotic birds with bright colours and in this case a silver white with black lacing. The skull is raised and on this part the rather large crest grows. However, an open drinker would result in the crest getting wet and bedraggled so use a drinking fount with narrow channel.

PRACTICAL HOUSING

The types of houses to use are many and varied, as shown previously. Irrespective of the type it will be necessary to consider the following:

1. Access to Shed and Run

The accommodation must be accessible so try to avoid very low sheds and runs which make it impossible for the poultry keeper to get inside. In addition, when there are a number of birds kept, have an outside run which has a door in which a wheelbarrow can be taken for dumping litter, such as weeds and taken from the garden and, when needed, to remove the accumulated manure.

2. Protection From Predators

There should be a 'pop hole' in the form of a sliding door, wide enough for the birds to go through. This must be closed each evening; or the outside pen must be fox-proof. This means having wire over the top of the run and putting wire netting into the ground within a 0.50 metre perimeter trench (fillled with stones and rubble), thus avoiding any possibility of animals digging underneath.

3. Have Padlocks on Doors

Valuable birds should be *kept* locked up at night because sometimes those who want a free meal, will take hens with no conscience, not realizing that the strain may have taken years to develop. Fortunately most fanciers are honest and would not steal. Old English Game are very tempting for the illegal cock fighting fraternity.

4. Adequate Ventilation Without Draughts

This is essential to allow birds fresh air, dry the manure and avoid the ammonia smell. Poultry must have lots of fresh air or they will not be healthy. They cannot stand dampness or excessive draughts.

Yokohamas

These are to be found in large and bantams. They are related to the
Japanese Long Tailed fowl which have really long tails, which reach
almost 10 metres. The Yokahamas lay tinted eggs and are worth keeping,   provided there is suitable accommodation to make sure they are
not running on heavy, muddy ground which spoils the plumage.

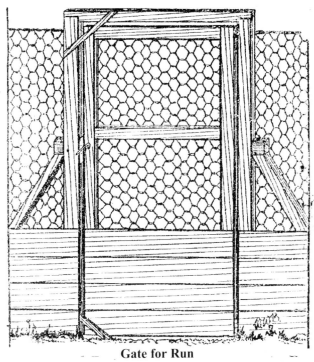

Gate for Run

This should be wide eniough to get a wheel barrow through. Weeds and leaves can be tipped from the garden. Also manure can be taken out periodically.

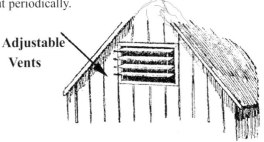

Adjustable Vents

One Method of Ventilation

Windows which open inwards are also effective, but wire netting should be placed over the gap or intruders may climb in, such as cats or foxes.

Black Rosecomb

Scots Grey

Partridge
Wyandotte

Silver Sebright

Four Favourite Bantam Hens

Head of Rosecomb Cock
Comb, showing the detailed
'workings' on the top with a
spike or leader at the back.
Besides the large ear lobe
the male has very wide
feathers in its full tail.

Japanese Bantams
Small, dwarf-like bantams which are kept as show birds.

Rosecomb Bantams
Very old breed of bantams, jaunty and attractive. Its main features are
its attractive plumage, detailed comb and ear lobes. It is found in Black,
Blue and White, but other colours do exist.

FEEDING THE BEST

Hens can be fed all the household scraps and will love them, but for maxmum production remember that these should be limited to not more than 20 per cent of the total intake. Thus bread, soaked in water or spare milk, cheese, rice, meat scraps and other tit-bits will all be consumed and the birds will become quite tame when being fed.

However, *do not* feed mouldy bread or contaminated food; this may cause problems and taint the eggs. Remember also that if an overall (average) protein level is to be achieved any bulk food, such as bread, should be balanced with the addition of bacon scraps, minced meat, or soya, thus bringing up the average to around 20 per cent. Meal, known as 'Mash' can be purchased and the household scraps can be mixed with this powder, using water or spare milk, thus giving a balanced ration.

Fowl have a feeding cycle which is along the following lines:

1. The Sitting Hen

She eats little food when she is sitting and provided she has corn and water each day is healthier from sitting. Give mixed corn with plenty of maize and limited water. Nothing that is likely to cause loose bowels should be fed.

2. Chickenhood

The small chick eats nothing the first two days and then gradually increases the amount, but in the first two months casts very little to develop. However, the food must be sound and contain sufficient protein.Do not hope to rear on scraps because this will not pay. Also chick crumbs from the local mill will contain a medicine which protects against coccidiosis, a condition caused by a germ picked up by the chicks running outside. Generally, though chicks are healthy and require little attention other than than good food, adequate fresh water and warmth.

If spare cockerels are to be fattened this is done for about 2 months; ie, to kill at 4 montha Extra food of the high protein type is essential.

4. Gradual Growing Period for Youngsters
Pullets should be brought on gradually and must not be over fed It has been found that if fed too much the pullets will start to lay at 5/6 months of age and the eggs will be too small. Therefore avoid over feeding and limit the protein Growers' pellets are suitable.

Pullets may be purchased about 5 months old and placed in a pre-prepared shed and run. This limits space required and replacements can be obtained every two years. However, if showing it will be desirable to breed your own 'winners' because this is the only way to establish a winning strain.

Will the Hobby Pay?
Keeping poultry does pay and there is a great deal of satisfaction from keeping and breeding birds. If all the costs are considered it should be found that even with very basic records kept they will show a profit. If a household uses two dozen eggs this is an immediate saving which offsets the budget. However, poultry keeping as a hobby has great therapeutic value which cannot be measured; it is relaxing, interesting and so worthwhile,

Matters to be taken into account are as follows:

1. Cost of Shed and Run.
This may be a converted shed or one purchased specially. It should last for many years so the annual depreciation will be a small amount.

2 Cost of Stock
These can be purchased at different ages and the younger they are the cheaper they shoud be. However, when first starting it is wise to obtain growing stock, fully feathered.

3. Food Costs

Food costs per bird on an average input of, say, 4 - 5 oz,per bird This may have to he varied depending on the breed.

4. Keep Records of Eggs and Cockerels Supplied to Outside and Household

In addition, if showing and winning there will be a ready sale of stock birds to those wanting new blood

5. Keep A Record of Manure Obtained for Garden

There will also be the poultry manure supplied for the garden which is very strong and used properly (after drying) is quite effective.

The value of manure will depend on a variety of factors but generally poultry manure is regarded as 12 times stronger than manure from cows. Its problem is excessive phosphoric acid which leaches plants, including grass. Potash should be added to off -balance, the phosphoric acid effect.

One mature bird will produce around 90lb. of manure, half to be collected from the shed droppings board. It is about 60 per cent water; 1.50 per cent nitrogen and around 0.75 per cent phosphoric acid. If peat moss is spread on the droppings board this should help to dry the manure.

Once collected the manure is placed in a covered area such as a compost shed and when dried thoroughly, is used as a fertilizer, but not too generously. Once dried the percentages of the fertilizer are increased to a higher level; eg, kiln dried manure has over 4 per cent nitrogen.

Probably on a small scale, if an *outside* person had to be employed, it would be difficult to get the poultry-keeping to pay, but not counting the "cost" of free labour given by the amateur, there should be no problem. There are however many additional benefits from a hobby which pays for itself. Knowing that eggs are fresh and the odd cockerel provides that extra meal when needed are incentives in themselves.

11

WELFARE RECOMMENDATIONS
Recent Trends

There have been many changes in the attitude towards the welfare of poultry, especially in relation to keeping birds in cages in overcrowded conditions, four birds to a cage, when they can hardly move. After about 50 years the conscience of those that make decisions has been pricked by the 'discovery' that birds suffered brittle bones, respiratory disease and sore feet as a result of this imprisonment.

Fortunately, garden poultry keeping does not subject the birds to excesses and the legislation or Codes* do not generally apply to the hobby. Nevertheless, it is important to remember that there is still an obligation to stick to standards which can invoke no criticism. As applied to the amateur poultry keeper the following apply:

1. Give Ample Food of the correct quality and an ever ready supply of fresh water.

Pools of muddy water are a source of germs so these should be eliminated so the birds cannot drink from them.

Generally birds consume about 5 per cent of their body weight and the exact amount depends on the energy supplied in the specific food.

2. Avoid Discomfort to the Birds.

They should have a dry water proof house with dry shavings or other litter on the floor. There should be regular cleaning of the poultry house. Ventilation should be adequate and poultry require plenty of fresh air but without draughts.

Codes of Recommendations on the welfare of Domestic Fowl, Ministry of Agriculture & Fisheries, 1987 and 1995. Also the recommendations of the Animal Welfare Council, RSPCA, Compassion in World Farming, and World Bantam & Poultry Society.

3. Watch for any signs of Diseases, Injuries or Discomforts.

This covers a whole range of possibilties and includes diseases and vices such as feather pecking. It also embraces mite and lice on birds which can be avoided by observing the rules of cleanliness and proving a dust bath and dusting the birds with insect powder.

Fortunately, the possible bad habits of feather pecking do not occur much in poultry kept on a small scale. If feathers start to look rather dishevelled then feather mite may be suspected. Alternatively, the cock may be too attentive and wears off the feathers on the back. If feather mite is suspected the birds should be washed in warm water including a mild disinfectant and this will kill any mite. The neck hackle is prone to this problem, especially on hard feathered birds such as Old English Game.

Birds with ailments should be spotted and isolated. The signs are standing with ruffled feathers, looking mopey, running eyes, not laying, pale face, faeces (droppings of a greenish colour) and any injuries. Usually drugs can be obtained to clear up simple problems, but any serious disease or injury would require the affected bird to be killed and incinerated. Fortunately poultry kept properly do not suffer much with diseases, but it is necessary to be aware of the possibility.*

Chicks can be a problem so they must be fed high quality chick crumbs with a Coccidiostat drug included, because Coccidiosis is the worst danger, when they are listless, lose their appetites, and do not develop in the normal way. Often the feather growth is retarded.

If the disease does hit them add medication to the water and observe strict cleanliness by clearing away the droppings on a daily basis. Many fanciers rear indoors in wire-fronted, wooden rearing hutches and this can be quite successful.

A few typical problems are shown opposite and generally these can be dealt with as they occur. With good management they may be avoided !

---

* **See Poultry** *Ailments for the Fancier,* **Joseph Batty, BPH**

Prolapse: may be due to
very large egg. Bathe and
apply ointment.

Crop Bound Bird
From eating unsuitable
food. Long grass or
excessive dry feeding.

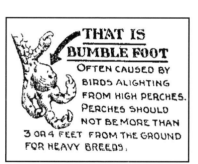

THAT IS BUMBLE FOOT
OFTEN CAUSED BY
BIRDS ALIGHTING
FROM HIGH PERCHES.
PERCHES SHOULD
NOT BE MORE THAN
3 OR 4 FEET FROM THE GROUND
FOR HEAVY BREEDS.

FROST-BITES

LARGE COMBED POULTRY
ARE SUBJECT TO FROST-BITE
THE COMBS GO WHITE, RUB
THE COMB WITH SNOW, OR COLD
WATER TO RESTORE CIRCULATION.

PURE VASELINE
RUB ON WITH A
PIECE OF SPONGE
OR FINGERS

PROTECT
ALL LARGE COMBED BIRDS
BY RUBBING ON VASELINE
EVERY FEW DAYS.

4. Provision of Natural Means for Normal Living

Poultry are scratching birds which forage for a great deal of their food. They should be able to run around, especially outside where there is grass or soil. They can pick up grit, soil and other requirements and enjoy the sun on their backs. Keeping birds in a very small space so they cannot flap their wings or turn round in comfort is quite wrong.

5. Avoidance of Conditions which Cause Distress

Making loud noises, chasing the birds around, allowing them to fight and bully and any other factors which cause likely distress should be avoided. Broody hens can be frightened by loud noises so care should be taken not to disturb them.

When birds are being shown they should not be left in a hot car or even in a show tent for long periods on hot days. Some of the two day shows are really too long.

If birds are kept in a Penning Room for training this should only be for short periods of, say, a week, or they start to lose condition and nails grow rather long, which have to be trimmed. Take care to trim below the line of the veins or the cutting will cause the bird to bleed.

Utilize the Old Shed at the Bottom of the Garden ?

INDEX

A Thought for A Useful Bird

The domestic fowl supplies us with eggs for break-
fast, with roast chicken later in the day, with feathers
to stuff our pillows, adorn ladies' hats and to make
flies to catch fish; indeed, we can truly say, where
should we be without her?

(Frances Pitt)